9
5795
983
c. 2

# Standards
and Standardization

# Standards and Standardization

Basic Principles and Applications

*CHARLES D. SULLIVAN*
*Manager, Technical Standards*
*Tektronix, Inc.*
*Beaverton, Oregon*

MARCEL DEKKER, INC.   New York and Basel

Library of Congress Cataloging in Publication Data

Sullivan, Charles D., [date]
   Standards and standardization.

   Includes index.
   1. Standardization.  I.  Title.
T59.S795    1983    389'.6    83-1884
ISBN  0-8247-1919-0

MARCEL DEKKER, INC.
270 Madison Avenue, New York, New York  10016

Current printing (last digit):
10  9  8  7  6  5  4  3  2  1

PRINTED IN THE UNITED STATES OF AMERICA

## PREFACE

The material presented in this book is a compilation of public information from various sources gleaned during many years of on-the-job involvement with standards and related subjects.  The author has tried to provide a broad coverage of standards so that the reader will have a comprehensive overview of the broad field encompassed by this subject.

Although an earnest effort has been made to present accurate information, the volatile nature of standards organizations, both domestic and international, may already have resulted in changes that differ from the text.  The reader is encouraged to contact any organization directly for more specific information.  Addresses are provided in the appendixes.

Charles D. Sullivan

*This book is dedicated*
*to the fond memory of*
Charles D. Sullivan
(1916-1983)

# CONTENTS

# INTRODUCTION

There is a growing awareness that standards have a part to play in education. Many standards, of course, do serve educational purposes, and there has been some use of standards in technological courses (see 12.3.5, 12.3.10). By and large, however, the knowledge and use of standards in general education has not been adequately stressed.

Standards are becoming increasingly important as a means of mitigating some of the problems of our modern society. They are also an important means of bridging many communication gaps.

Standards have many values, and may mean many things, depending on the use and the user. Mr. F. A. Sweet, formerly with the Canadian Standards Association, expressed four values of standards as follows:

First, they educate. They set forth ideals, or quality goals, for the guidance of manufacturers and users alike. They are invaluable to the manufacturer who wishes to enter a new field and to the naive purchaser who wants to by a new product.

Secondly, they simplify. They reduce the number of sizes, the variety of processes, the amount of stock, and the paperwork that largely accounts for the overhead costs of making and selling.

Third, they conserve. By making possible large-scale production of standard designs, they encourage better tooling, more careful design, and more precise controls, and thereby reduce the production of defective and surplus pieces.

In so doing, of course, they also benefit the user through lower costs. Our present-day standard of living would not be possible if we had not achieved parts standardization. Now, many large, divisionalized companies are emphasizing control and standardization of parts and components throughout the total organization, so one division will not be throwing away parts another division may be purchasing on the market. This is all handled by computer programs and a bar code that identifies products in much the same way as the checkout at the grocery store.

Fourth, they are a base upon which to certify. They serve as hallmarks of quality which are of inestimable value to the advertiser who points to proven values, and to the buyer who sees the accredited trademark, nameplate, or label.

Sometimes we are not all that happy with the quality we get, even with such certification, but we must remember the vastness of the marketplace and the impossibility of policing every transaction. Indeed, the fact that our system works at all is due more to the basic honesty of people than to regulations.

Properly conceived and used, standards can help people cope with the increasing technical complexity of our times.

Standards are essential tools in the interaction of people with their environment. They are essential as a means of assuring consumer protection in the use of the myriad products that constantly increase our choice and selection of what we have come to regard as the necessities of modern living.

The interpretive factor for understanding and observing regulative controls is best expressed in the form of standards. Even in disciplines not directly connected with engineering and science, some understanding of the basic nature of standards is becoming increasingly relevant.

For example, an economist must have a knowledge of standards and how they originate in order to recognize their effect on production and trade and their influence on the international exchange of goods and services. A lawyer must be aware of standards as instruments dealing with product liability, safety, and environment—and, of course, contracts. With the intrusion of technology into medicine, doctors are finding that a knowledge of standards can be of great help in dealing with instruments and methods to protect their patients. Hospital administrators know they must be aware of various federal standards and abide by them or their hospitals can be denied federal certification.

As life becomes more complex, more controls are needed. Standards are the documents that carry these controls throughout the social structure. The nature and significance of standards, however, need to be more widely recognized. We are all at least vaguely aware of the vast structure of laws that has been created since American independence from Great Britain. Indeed, it might be reasonable to ask, "what independence?" Still, we manage to function with some measure of personal happiness in spite of this vast body of law, and we can even admit, when we stop to think about it, that these laws are generally beneficial.

It is not necessary that everyone affected by standards understand what every standard says. Just as it is not possible to know all the laws, only a few specialists generally understand the more technical standards. However, as in the case of public law, a general knowledge of what standards are, how they are used, and what their impact can be, is a valuable asset in any vocation.

Strangely, in almost all countries there is a surprising lack of information on standards in the schools, even in countries where technology is highly advanced. The number of teachers of standards is also exceedingly restricted since most of the knowledgeable standards people are in private industry or government, where they have had to acquire the knowledge as they progressed in their particular occupations. If schools do not teach students about standards, the academic awareness that provides for the teaching of standards will not be developed. There is a gross need for more knowledge of standards and their significance.

These things being so, I have made an effort here to present some of the basic information on standards documents and practices so that students will be aware of, and better equipped to deal with, standards in whatever vocation they select.

Since all standards documents are prepared in a format which uses a peculiar paragraphing and indexing, I have chosen to write this book in the form of a standard document. Although it is not possible to totally follow such a format, I feel this style will more effectively present the subject than the more traditional text approach. In some formats, paragraphs are referred to as sections, or clauses, but whatever the term, the object is to provide easy identification of subject matter and cross-reference.

# Standards
# and Standardization

# 1. PURPOSE

It is the purpose of this volume to introduce readers to standards and to provide information which will assist them in not only understanding such documents, but also the reason for standards and the influence they have on us and the world in which we live. In a secondary sense, it is expected that anyone who must use standards will benefit from a knowledge of their nature and how they are formed.

## 2. DEFINITION OF A STANDARD

There have been many heated arguments over the definition of a standard. In this text, however, I will treat standards as a category of documents whose function is to control some aspect of human endeavor. It is an exceedingly large field.

### 2.1 Approaches to Standards

There are two basic approaches to standards—active and reactive documentation. These may be described as follows:

### 2.1.1 Active Standards

Active standards are planned standards, resulting from forethought as to their need and content. For example, a Mr. Ford invents a mechanical wagon that can run by itself. The inventor realizes that when this contraption appears in public it will cause considerable problems in traffic, particularly in regard to roads and vehicles pulled by horses. Mr. Ford accordingly gathers a group of public officials and explains to them the need to establish some traffic standards that will help to avoid problems. This is the active approach to standards; the new vehicle and new controls appear at the same time.

### 2.1.2 Reactive Standards

In the case of Mr. Ford and his horseless carriage, if we imagine that neither Mr. Ford or any of the people involved happened to think about the effects, the havoc the first vehicle created on its first drive down the main street would quickly reveal the need for some controls. These would, of course, be reactive in nature.

While this description is a little on the historical side, similar situations and opportunities are everywhere around us. With every new machine, instrument, or process, there is the opportunity to plan for standards that will have to be written. Certainly it is much less costly to plan and implement standards along with a new product or process than to have to write standards later to solve problems created by the new introduction.

## 3.  SCOPE

Standards, in both written and unwritten form, control conditions, aspects, or behavior of practically everything on earth.  There are few human-use objects or activities that are not influenced or controlled by standards.

### 3.1  A Succession of Controls

Standards represent a succession of controls paralleling the growth and development of the individual.  The first controls appear in the family.

#### 3.1.1  The Family

From the cave to the condominium, and from the cradle to at least the first grade, the mores of the family have exerted control over the individual.  In these early years the control is parental, but it can also include other family members, relatives, the doctor, minister, and others intimately connected to the family group.  As the individual grows, the enviroment expands outside the family.

#### 3.1.2  Society

Society, as as extension of the family, becomes a control factor at the beginning of our school years.  Thereafter, societal control rapidly increases.  Usually we are not aware of these increasing controls, and almost never are we aware of the standards on which these controls are based.  Finally, we begin our working lives and more controls affect our activities.

#### 3.1.3  The Working World

As we mature and begin our working lives we come into contact with additional controls.  Some are established by our employer for hours and conditions of work, others are associated with a particular function.  Again, we seldom are conscious of the documents behind the controls.

Controls based on voluntary standards are usually beneficial to the individual and to society. Controls created by edict can often be cruel and inhuman. We will deal primarily with voluntary consensus standards in this text (see paragraph 11.4.2).

# 4. STANDARDS ARE ANCIENT THINGS

The controls mentioned in paragraphs 3.1.1, 3.1.2, and 3.1.3 are primarily spontaneous and traditional, a part of our physical growth and development. They are not usually backed by written documents. On the other hand when people consciously get together to develop a method to solve a recurring problem or establish standards by plan in order to avoid the problem, as explained in 2.1.1, they create intentional controls (standards) that require written form.

## 4.1 Earliest Written Standards

The earliest written standards were for weights and measures—to control dealings between individuals. Some 5000 years ago the Egyptians developed a standard of measurement based on the length of the pharaoh's forearm. It was approximately 20.63 inches long, and was divided into 6-palms and 24-finger widths inscribed on a block of black granite. This standard, called the Egyptian royal cubit, was established as the basic measure for the pyramids and other great monuments. As in our modern National Bureau of Standards, this master measuring instrument was carefully protected and was not itself used as a tool. Wood copies were made to be used where needed for construction projects. As an exercise, if you will measure a few forearms, you will see why just specifying "forearm" would not do; it had to be a specific (identified) forearm length, and the pharaoh got the honor.

## 4.2 Even the Ark

In Genesis 6:11, God instructed Noah to build the ark, measuring it in cubits. It was to be 300 cubits in length, 50 cubits in width, and 30 cubits in height. I do not know whether these cubits were based on the royal cubit of the Egyptians.

## 4.3 Historical Highlights

There have been many historical highlights in standards. In one book on the subject*, a description is given of the measurement systems of the ancient Middle East—Egypt, Phoenicia, and Babylon. These systems were oriented to the purposes of peace and commerce, and formed a substantially uniform and rational decimal system somewhat similar to today's metric system. However, Greece introduced the pace and the foot into this system for military purposes. Eventually, as various civilizations succeeded each other, the measurement system fragmented. Today the world is trying to reestablish a single rationalized metric system known as Systems International (SI).

Due to the nature of this book, it is not relevant to give extensive coverage to SI. However, SI is a rationalized selection of units from earlier "metric" systems. It is a coherent system with seven base units that have precise definitions and symbols. From these, many derived units have been established. The system, like our monetary system, is based on 10 and its powers. Very likely, the whole decimal structure rests on the anthropological fact that humans have ten fingers and ten toes.

There are still many national measurement systems, either resisting the SI, or paralleling its use, but the change appears to be inevitable.

The metric system of which Erwin spoke as coming to dominate the world has itself undergone many changes based on international needs. It is certain that worldwide adoption of SI will require many changes in existing documents. The need for change may be triggered by technological development, obsolescence of the subject for which the standard was written, new laws, new governments, and so on. It is essential that standards be reviewed frequently to determine whether they are still valid.

This review is itself a huge task when we consider how many thousands of standards there are, and the demands on the time of people involved with the new standards. Although revisions are often a hindrance to new work, someone has to keep these standards current.

### 4.3.1. A Reason for Standards

It is interesting that the earliest controls (standards) were originated to "keep people honest." Since the dawn of human society, goods and services have been exchanged between people on the basis of physical measurement, and from such equally ancient times people have cheated and shortchanged their neighbors. Like the person flipping the two-headed coin, unfair dealings and false measurement have been resented, and systems have been established to enforce honesty.

---

*K. G. Erwin, *Romance of Weights and Measures* (New York: Viking Press, 1960).

*4.3.1.1 Effectiveness of Standards.* The mere establishment of a standard to control behavior won't, of itself, keep everyone honest. There must be sufficient value in it, and a sufficient number of people willing to observe it, or the standard will not be effective. For those who do not want to abide by standards, and seek an unfair advantage, there are consequences (see paragraphs 5.1 and 5.2).

However, practically any standard, even though mandatory, may be "fractured" for good and sufficient cause. The reason is that it is not possible to either perfectly define all possible uses, nor to foresee all contingencies. This in no way detracts from the authority of the document.

# 5. TYPES OF STANDARDS DOCUMENTS

In paragraph 2.1, we indentified active and reactive standards. Standards can be written and unwritten, voluntary or mandatory.

(*Note*: For ease of writing and reading, the various kinds of standards documents will be referred to in the following pages as simply "standards." A closer distinction will be made only where necessary.)

## 5.1 Voluntary Standards

Although it would seem that voluntary standards ought to be standards which can be either used or rejected, this does not generally hold true in practice. Our society places a high value on conformity; as soon as a number of people come to accept a voluntary standard they expect others to conform. Individuals (or companies, etc.) that insist on doing things their own way may find themselves at an increasing disadvantage.

For example, suppose that several companies are making similar electronic products. Because it is not economically feasible to make every part they need themselves, they go to a vendor. The vendor sees how a part can be made that will satisfy the needs of all the companies, thus reducing the tooling cost and offering the manufacturers the benefit of cheaper parts. However, one company wants to stick to its own design. The vendor says that the special part can't be made right away because of the demand for the standard part. That leaves the manufacturer the choice of continuing to make the special part at higher cost or accepting the standard part. A hurt pocketbook can be a powerful force for observance of standards.

There are many good reasons why voluntary standards ought to be considered as something a little stronger than "voluntary." Although they are created by voluntary action of some group, there would be no reason for the effort and cost of developing the standard unless it was expected that people who used the document would benefit by doing so.

## 5.2 Mandatory Standards

Essentially, mandatory standards are laws. Unlike the case of voluntary standards, which may or may not have consequences for nonobservance, failure to obey laws will invoke legal sanctions and penalties. Many voluntary standards evolve into mandatory standards on the theory of the greatest good for the greatest number. The individuals (or companies, etc.) come under protection whether they wish it or not. It is not easy to take over a voluntary standard and use it for mandatory purposes. This situation existed when Congress created the Occupational Safety and Health Administration (OSHA). This act provided for a government body on which to base enforcement. OSHA had to have standards in a hurry, so it adopted many voluntary standards that had been developed by industry. This was not especially effective since the language of these standards was liberally laced with "shoulds." It took some time to get some good OSHA standards.

(*Note*: A curious fact is that most standards are created, approved, and implemented, not by majorities, but by minorities. See paragraph 11.6.2.3.)

## 6. USES OF STANDARDS

In paragraph 2, standards were defined as a category of control doc-
ments. We will develop this concept as we proceed. Some of the things
controlled are quantity, extent, quality, value, and methods or activities. When considered in relation to quality and value, standards assure that we will get at least reasonably fair value for our money. We
do not have to personally double-check on everything we buy, because
we know (although we generally don't think about it) that the product
has been developed according to standards to assure good quality and
value. Unfortunately, there always will be a degree of dishonesty in
any dealing, but without standards it would be far worse.

# 7. SUBCATEGORIES

In a general introduction to standards such as this book intends, it is not possible to cover all the subcategories of standards. There are a great many kinds of standards, and the following section will describe only the more prominent ones with which we come into contact.

## 7.1 International

Until fairly recently, even people whose work involved standards were not especially aware of international involvement. However, the Second World War showed the need for coordination of such seemingly simple things as fasteners, and lent a very real emphasis to the effort to co-ordinate standards between allied nations (see paragraph 14). Since that time, various treaties and trade agreements, and especially the creation of the European Common Market (ECM), have increased this emphasis. Now, third-world countries, as they emerge onto the world scene, are demanding standards that will allow them to produce and market their goods on a global scale. What they are precisely demanding, stated in another way, is standards that do not discriminate against them.

(Standards can indeed discriminate. In early days, many trade unions and guilds intentionally wrote standards that outsiders could not meet, thereby stifling competition. Today, while intentional discrimination is rare, standards written by technically advanced nations may contain conditions that less developed nations do not have the facilities to meet, and thus prevent them from competing on the world market.)

Of course, this effort for nondiscrimination can be carried just so far. You cannot, for instance, put off writing standards for automobiles because some new country may still use camels; nor can you jeopardize public health in your own country by loosening sanitation controls on foodstuffs so that a less advanced country can market products grown and handled under less sanitary conditions. What happens is that the advanced countries build automobile factories in the camel pastures, and sell sanitary facilities to the less advanced nation.

So the pace of international standardization is rapidly accelerating. Among the principal international standards organizations are the International Organization for Standardization (ISO) and the International Electrotechnical Commission (IEC). Each will be discussed later.

(*Note*: There are hundreds of organizations outside the electrical and mechanical engineering fields, including international groups such as the World Health Organization [WHO], that are involved with standards. It would require a library of books to describe every organiza organization.)

## 7.2 National

Except in the United States (see paragraph 7.2.3), most nations have developed a national policy towards standards. In some countries, state funds have been allotted to the standards effort. In the Soviet Union, there is a minister of standards. For further discussion, see paragraph 15.

### 7.2.1 Voluntary Standards Organizations in the United States

There are a great many organizations in the United States that produce standards. From various sources I have heard that there are more than 400 such organizations. An excellent source of information is the *United States Standardization Activities*, published by NBS. Information on many associations can also be obtained by consulting *An Encyclopedia of Associations*, published by Gale Research Company. It lists thousands of organizations, including the standards-producing associations, but these are not separately identified. This Encyclopedia provides addresses and brief sketches of the organizations and is a very informative source. A similar publication is the *Directory of National Trade and Professional Associations of the United States and Canada* and *Labor Unions*, published by Columbia Book, Inc. Americans, it seems, have associations for everything from chocolate cake to little league baseball. For reader convenience, I have included a selected list with complete addresses in App. D.

*7.2.1.1 Major U.S. Standards Developing Organizations.* The American Society for Testing and Materials (ASTM), American Society of Mechanical Engineers (ASME), Society of Automotive Engineers (SAE), Aerospace Industries Association (AIA), National Fire Protection Association (NFPA), and the American Society for Heating, Refrigeration, and Air Conditioning Engineers (ASHRAE) are generally considered to be the largest standards developing organizations in the United States. Underwriters' Laboratories, Inc. (UL, see paragraph 12.10) and the American National Standards Institute (ANSI, see paragraph 12.1), are popularly considered to be producers of voluntary standards, but ANSI actually functions more as a custodian, coordinator, and accreditor of standards, while UL deals with testing (safety) and listing of products which pass these tests. UL does, however, write product safety standards. Addresses for all of these organizations can be found in App. D.

*7. 2. 1. 2 International Scope of Standards Organizations.* The U.S. voluntary standards organizations have conducted their work primarily in response to the needs of U.S. industry. However, some have correlated their work with other nations and have also established membership on international committees involving other nations of the world (see paragraph 14).

*7. 2. 1. 3 Government Participation in Voluntary Standards Organization.* Although funds have not been made available to U.S. standards organizations by the federal government, the organizations have cooperated with federal agencies and the Department of Defense (DOD) in the development of standards. Many military and other government representatives have membership on the committees of the voluntary standards organizations. In an increasing number of cases military and federal agencies have adopted standards produced by these organizations. The Office of Management and Budget (OMB), in order to facilitate government interaction with the voluntary standards organizations, issued a circular called A-119 to establish conditions for federal participation in voluntary standards work. The purpose of this circular is to establish a policy to be followed by executive branch agencies in working with organizations that produce or coordinate voluntary standards for materials, product systems, services, or practices. It was also intended that the circular would establish a policy to be followed by executive branch agencies in adopting and using such standards in procurement activities.

*7. 2. 1. 4 Criteria for Government Participation.* This statement of policy really does not tell much of the story. In order for any standards organization to benefit from and secure the participation of the government, such organizations must establish procedures approved by the government. Such procedures involve the basic principles of due process, wherein development work in standards is widely publicized, and the people writing the standard are from many disciplines.

*7. 2. 1. 5 Objective of Government Criteria.* The objective is to avoid injury to any interested persons through lack of knowledge of what is being promulgated. In very plain language, the governement would not do business with any organizations that didn't get their act together and meet the government's criteria for participation. This in no way means the majority of the voluntary standards agencies were not already following the precepts of due process—just that the "feds" wanted a document that said what the rules were. At this time the OMB circular has not yet been approved.

*7.2.1.6 OMB "Standards."* The full text of the OMB "standard" is much too detailed to include here, but it met with the general approval of the major standards bodies. Additional information may be obtained by writing to National Bureau of Standards.

### 7.2.2 Government Takeover?

At various times, bills have been introduced in Congress to provide for government control of voluntary standards-making organizations through the office of the Secretary of Commerce. Such bills do not have the approval of the standards community. Instead, these organizations encourage increased federal agency involvement in creating necessary standards.

### 7.2.3 National Standards Policy

In 1977 a recommendation was made by the National Standards Policy Advisory Committee (NSPAC, see App. D for address) for a national standard policy. The American National Standards Institute acted as the guiding element in drawing up this proposed policy, after which it was released for public comment.

The document contained a recommendation to establish a private sector standards coordinating center, and ANSI, as the already accepted "coordinator" for voluntary standards in the United States, seemed the logical organization to fill this role. It required a neutral group (i.e., a nonstandards-writing group) that would represent the position and interests of the United States, and would serve as the recognized agency in this country for participaion in international standards bodies.

However, some organizations could not see ANSI as successfully fulfilling this role without making some basic changes in its structure, and at this writing a national standards policy is still not official. Eventually, of course, such a policy will have to be established if American industry and government are to have an effective voice in international standards.

As an example of the complexity of standards interaction, ANSI in 1980 drew up a plan for restructuring its international standards functions and interaction with the U.S. government. This plan was put out for public review and comment in January 1981. Very briefly, it provided for a new charter for ANSI's International Standards Council (ISC) and a new U.S. Industry/Government Standards Affair Committee. The main thrust of the plan was to enable ANSI to retain its long established position as coordinator of the voluntary consensus standards system in the U.S. and to further the development of international standards that qualify for worldwide acceptance. Because of various factors involving both government and industry, the ANSI restructuring plan has not yet been fully implemented.

(*Note*: I have included this coverage on the national standards policy in the hope that it will provide a fuller understanding of the "big board" activity of standards. The reader should understand that this is being written in the midst of significant changes in the standards system. If the writing were to be postponed until everything was neatly embedded in concrete, there would not be any book to read.)

## 7.3 Industrial Standards

There are hundreds of industrial associations in the United States which have established committees and written standards for their own particular needs. The fact that such cooperative endeavor has been voluntarily entered into by competing companies should be a strong argument against the need for strong government control. Manufacturers have recognized the need for common application of methods and means for manufacturing, shipping, and marketing, and have gone ahead and written the necessary standards. Since the objective of business—in fact, the necessity—is to make a profit, it is clear that these businesses expected the standards to be a good investment.

### 7.3.1 Unfortunate Aspects

In paragraph 4.3.1, I commented on some of the reasons for the implementation of standards. In practically all phases of human activity, there is a self-interest factor that can be either selfish or altruistic. Occasionally it leads people to take advantage of situations for their own benefit. In the early days of America, there were instances where local standards were written to protect established trades and exclude competitive products that might be better or lowerpriced.

Federal laws for fair trade and due process pretty well corrected the situation as it applied to commercial standards, but on the international scene it was recognized that some national standards could be restrictive when it came to trade between nations (see paragraph 7.1). The General Agreements on Tariff and Trade (GATT) code is a present-day effort between nations to assure that such standards will not be written. (See App. B.)

## 7.4 Proprietary Company Standards

In spite of the advantage of similar methods of manufacturing, shipping, and marketing, every company develops certain ways of solving problems which, as with individuals, become habits. Like habits, they are not easily changed, and they may become "cast in concrete" by in-house procedures and specifications.

### 7.4.1 Dangers of Outdated Procedures

This situation is not all bad because habits are great time savers and as such are profitable to a company. The danger is that, since nothing in industry remains static, a company too addicted to peculiar ways may fall behind new competitors who are able to benefit by more modern approaches. An out-of-date practice is just as likely to be a liability as an asset.

### 7.4.2 Competitive Policing

An interesting fact is that when one company puts out a new product and states its specifications, very often one of the first purchases, and the first external test of the product, is by its foremost competitor.

It is possible that this competitor will find the new product actually does not comply with, say, some federal regulation, and consequently does not meet the published specifications. In such a case, it might be thought the competitor would bring this immediately to the attention of the regulatory agency and thus discredit its manufacturer. While it is difficult to generalize, probably the most prevalent action is for the discoverer of the defect to communicate directly with the manufacturer. There is heavy peer influence among companies, even competitors, and it is generally to the mutual benefit of the various manufacturers to conduct an internal policing operation, rather than to "run screaming" to the concerned regulatory agency.

### 7.5 Personal Standards

Although in this text we are of course more interested in higher and more complex systems of controls, the standards that control a company or an industry are the product of human interaction. They exert an influence on the nation, and even further, on relationships between nations.

We can see in our own daily routine many controls which we have voluntarily accepted (or adopted) and which affect our personal behavior and attitudes. On a larger scale, this routine applies to more complex categories of controls. Standards are, after all, the product of people trying to avoid problems by forethought or to solve them (see paragraph 2.1.1).

## 8. APPLICATION OF VOLUNTARY STANDARDS

The field of standardization is infinitely complex, and it is difficult to discuss any portion of it without encroaching on other areas. In general, voluntary standards in industry exert a broad control over materials, methods, and products, while allowing considerable flexibility as to the manner of application.

## 9. APPLICATION OF MANDATORY STANDARDS

Mandatory standards, as mentioned in paragraph 5.2, are obligatory. They are generally detailed, limited in scope, and very specialized. They are usually based on governmental controls, and a business concern has little choice but to observe the limitations placed on it if it wants to remain in operation.

### 9.1 Federal Standards

When mandatory standards have a legal base, such as the Occupational Safety and Health Act (OSHA), or the Environmental Protection Act (EPA), they are generally an explicit description of limits to which an industry must adhere for a product, material, service, or manufacturing operation. Under ideal circumstances, which are seldom attainable, these standards operate for the public good; in such areas as health, safety, and environmental protection.

Federal standards may specify supplies, test materials, methods, and processes. Material standards involve materials, products, and services oriented to the national defense. Information on most federal documents can be obtained from the U.S. Government Printing Office. (See App. D for complete address.)

# 10. STANDARDS FOR COMMERCIAL PRODUCTS

Commercial standards are established to provide a clear understanding between seller and buyer and may include type, grade, classification, size, and suitability of use. A common example is eggs. Why aren't they sold by the pound rather than by the dozen? When standards for eggs were established there had to be a decision on how to market them. Although most produce could be sold by weight, it was difficult to apply this to eggs. Since they are fragile, it was not practical to put them into a paper sack, so protective packaging had to be developed. Further, this packaging had to be uniform in size so that automatic equipment could be used, and even the eggs themselves had to be uniform in size. A great industry flourishes in the United States based on the egg, and the standards that had to be developed to make economic marketing possible. Although eggs are an agricultural product and therefore under the control of federal departments, this example shows how the nature of a product can dictate the development of directly associated standards.

(*Note*: Industrial standards, as a separate category, are concerned with the conduct of business, such as agreements between manufacturers on what constitutes fair practices in marketing their products. Many of these standards developed from "gentlemen's agreements" between competing companies in the early years of our country. Sadly, as commerce became more complex, "a handshake to seal the bargain" ceased to offer the necessary assurance. Nowadays people still shake hands as a symbolic affirmative, but the "shake" is now backed up by enough paper to have sunk the *Mayflower*.)

## 11. THE MECHANICS OF STANDARDS PREPARATION

Except for controls established by edict (and possibly not even then)
standards are the work of groups of people, generally identified as
"committees."

### 11.1 The Formation of Committees

Who decides a document is needed? Who decides a committee is needed?
Who decides what people will be on the committee?

It depends on the situation. A dictator seeking to secure a greater
control of the populace can arbitrarily select any group he pleases
and command them to provide a control document. A corporation can
assign to a specific person the responsibility for producing a standard
needed by the company. A governing body, such as a federal bureau,
can secure the services of anyone it believes has knowledge on a needed subject by inducement of pay or patriotism. Standards organizations can contact individuals through public announcement, or they can
contact industries involved with the subject and request their assistance. International organizations seek to secure capable individuals
by contacting national standards associations. By whatever means, a
group of people is formed to provide the needed document.

This process also takes place within most companies, although there
are a number of ways to create company standards, depending on how
a company standards program is set up. Generally, a company standards program is neither adequately funded or staffed (again, there are
differences), and few standards people could win popularity contests
for their occupation. That in no way lessens the importance or value
of standards to a company, and to most standards personnel the production of good standards is a real challenge, involving as it does the
need for maintaining equable relationships throughout all levels of the
company.

### 11.1.1 The Committee Members

Although there may be some members of a committee who are marginally
effective and who see committee work as prestige enhancement and a

chance to do some traveling at company expense, committee members are usually dedicated individuals who believe in the value of standards and therefore spend much of their own time and funds to further their development. The following editorial, which appeared in a publication of the American Society for Testing and Materials (ASTM), is a reprint of a piece written by the late A.Q. Mowbray, editor of the former *Materials Research and Standards* (now *Standardization News*). While the subject and the figures change, the dedication remains the same today as it did in Mowbray's time.

## NOT FOR HIRE

You are sitting in a meeting room in the Conrad Hilton Hotel in Chicago. It's getting late. Your hurried supper rests uneasily on a stomach tensed by tight schedules and fatigue. The air in the room is over-tired, smoke-laden, oxygen-depleted. Neckties are loosened and sleeves rolled up, but still the close-packed, sweaty faces of the audience are attentive, the voice of the speaker is strong.

These men were up until one or two a.m. the night before in a meeting of a task force or subcommittee, hammering out agreement on the details of a specification or test method, reporting test results, arguing, cajoling, listening. They were up again five hours later for another full day of the same. Between meetings they collar each other, cluster in hallways, impede traffic with the earnest, arm-waving accompaniment to their discourse. Meals are often a quick sandwich and a cup of coffee with the briefcase within arm's reach. If an opening appears in the schedule, they run to a session room to hear a paper or two that they had checked on their program.

Toward the end of the week, the fatigue lines begin to show. Why do they do it? Well, you hear a lot of reasons, all of them good. (1) They have to protect their company's interest in the standards-writing forum. (2) Working with these committees is good education, not only technically, but also in the art of democratic give-and-take, organizing research programs with one's peers. (3) Talking with authors and others at paper sessions is a good way to keep up with what is going on. And so on . . .

As you sit in this hot crowded room speculating on these things the presentation of the paper ends and questions begin to come from the floor. You are admiring the openness and freedom of inquiry and criticism inherent in this centuries-old format and reflecting on the self-correcting nature of scientific investigation, when, in answer to a question, the author replies, "Well, I was so pleased with the results (of a test) that I wanted to tell somebody." This statement shoots through

the stuffy room like a fresh breeze. You feel the laughter that
goes through the audience springs from the rare and pleasant
recognition of simple truth, simply spoken. For, after all the
other reasons are given, the root cause for all this effort is the
pleasure of the exertion. No truly good work was ever done re-
luctantly, or with distaste. This man, and dozens like him, had
done good work, and in the end, the reasons for having done it
were first, the enjoyment of doing good work, the second, the
pleasure of reporting it to others. Wherever this attitude appears, whether in the scientist, the carpenter, or the bus driv-
er, it is the truly professional attitude. No employer can buy
this dedication, no minimum wage can produce it, no incentive
plan can take its measure. (A. Q. Mowbray, 1964, © ASTM. Re-
printed/adapted with permission.)

*11.1.1.1 Personal Dedication of Participants.* I have been told altruism
today is like the dodo, nonexistent. The people who pay the bills
for the participants expect to get something for their money. If so,
the members may indeed just be "doing their job" and are protecting
the interests of the company by making certain views known in the
development of a standard. I believe we can accept this without in
any way detracting from the personal dedication of the individuals,
who often forego personal vacations with their families in order to
attend committee meetings. Without the incentive of being a part of
something good and worthwhile, there would be no committees of a voluntary nature.

*11.1.1.2 Participants Are Human Too.* One should not forget, however, that these are just ordinary people, although experts in a partic-
lar field, and sometimes in meetings arguments can become a little personal. Two experts can't both be right on opposite sides of an argument. It requires a comprehensive awareness and a high degree of mutual respect among the members of the committee to safely bridge these
moments of individual pique.

## 11.2 Committee Chairman

I use the word chairman strictly in a neutral sense as meaning any
person, male or female, who occupies the lead position on any committee. All committees must have a leader. This person may be no more
knowledgeable than the other members, but nevertheless has great influence on the character of the standard and the time required to produce it. The chairman's example sets the tempo for the total membership.

## 11.3 Committee Action

The committee may be considered a production team, with the standard as the product. No quality product can be made without good material and a sincere effort.

### 11.3.1 Source of Material

Generally, committee members are knowledgeable on the subject for which the standard is needed. It may seem strange that some members may not be experts, but various talents are required besides technical knowledge, including illustrating, writing, and editing. Regardless of their personal knowledge, committee members establish contacts and identify sources from which authoritative information can be obtained. In most cases, previous standards or workplace experience in the same, or related, fields provide basic information on which to build. Information may also come from academic or government research, such as that conducted by the National Bureau of Standards.

### 11.3.2 Sponsorship

Standards that are highly complex, or many pages in length, are generally sectionalized. In such cases, individual committee members may accept the responsibility for particular sections. This approach spreads the work around somewhat and makes for faster progress because each sponsor becomes a focal point for input and review pertaining to the subject of the section. Usually such individuals are recognized as being highly competent in that subject.

*11.3.2.1 Sometimes It Is Not Much Fun Being a Sponsor.* These committee members painstakingly put together the comments of the other members and circulate what they feel is a consensus view, only to sometimes have it torn completely apart. Again they must reassemble the pieces until they achieve what they consider a coverage no sensible person could find fault with. After a while they begin to doubt whether there are any reasonable people besides themselves on the committee. Nevertheless, the work goes on.

### 11.3.3 Meetings

Meetings are an essential part of the committee process. The time and place of meetings varies widely from small, in-company meeting rooms with a few individuals meeting perhaps once a week, to international groups who convene once a year, or even every two years. In particular cases where there are a large number of members it may not be practical for all members to meet at once. In such cases, subcommittees or work groups may be formed to carry on portions of the work and meet independently of the others. It is then usually necessary to have liaison persons to monitor the activities of the various groups.

*11. 3. 3. 1 National Committees.* Committeess established on a national basis generally meet several times a year in various parts of the country to equalize travel time and costs for the members. If the standards being worked on are of an industrial nature, the members are usually company employees and the cost of their participation is financed by their companies in the interest of standardizing conditions which affect company activities.

## 11.4 The First Draft

Like everything else, standards have to have a beginning. There is nothing easy about writing a standard. It is a genuine challenge to create order and lucidity out of the English language which is an admixture of many languages.

### 11.4.1 Amount of Work Needed

The amount of work needed to produce a draft standard varies with the amount of information available. Where there is little or no prior information, as might be the case with radioactive space debris, the work of assembling a document may be a long and difficult task. It may, also, be largely hypothetical until actual conditions can be measured and evaluated.

*11. 4. 1. 1 The First Effort.* This may be a very rough preliminary draft which is more a sounding board than a document. It is put together to provide a focal point for input and comment so that many people who have knowledge of the subject can provide information.

*11. 4. 1. 2 Objective of the Preliminary Draft.* Where there are previous standards or where considerable testing and experimenting has taken place, there may be sufficient material to put out a reasonably developed document. Regardless of the condition of the draft standard, it is still only a beginning. Its function is to present the subject to as large a number of people as is reasonably possible in order to get maximum exposure and input. Generally speaking, the larger the exposure, the better the standard will be. However, this is no hard and fast rule.

(*Note*: Because of federal rules and regulations, all voluntary standards-making bodies try to publicize standards projects so those who will be affected by the completed document will not only be aware of the standard, but will have the opportunity to make their views known while the standard is in preparation. [See paragraph 12.1.1.6.])

*11. 4. 1. 3 Rewrite and Rewrite.* Depending on many factors, some standards may go through a dozen drafts before they reach approval.

### 11.4.2 Consensus Agreement

After every circulation for public comment, a committee must evaluate the returned comments and judge what is good and what is not really pertinent to add to the document. The process goes on and on, with the standard being whetted a little more until it is considered by those involved to be as good as it can be under the circumstances. This is consensus agreement.

*11.4.2.1 There Are Very Few, If Any, Perfect Standards.* What comes out of the hard work is a document which in effect is a document known to be imperfect, yet acceptable at the time for the purpose for which it was made.

### 11.5 Completion

When it is finally decided that further effort would not produce sufficient benefit to justify continued work at the time, and would likely prolong completion to the point where the standard would not be timely enough to be fully useful, the draft work is considered to be completed. It is then time to seek approval of the document.

### 11.6 Approval

Approval for issue cannot be made by the members of the committee. Approval must come from the people who provided the information or, in some cases, from a higher committee which reviews the standard and exposes it to reviews by those who will likely be affected by its provisions. Ideally, all standards organizations strive for a consensus approval of their standards. The American National Standards Institute (ANSI) provides an independent verification service for consensus.

### 11.6.1 Methods of Approving Standards

There are three generally recognized methods of approving voluntary standards. The first is the canvass method, under which public notice is given of all development activity and the public is invited to make comments.

(*Note*: The word "public" as used here means that notice is made in various media which have public circulation; however, such media is generally the type that will be read by people whose interests are in that field, or in associated fields, and who will consequently have an interest in such standards.)

The second method is the accredited organization method. Standards are developed by organizations accredited by ANSI. These organizations are recognized as having high technical qualifications in the field of the standard (such as ASME in boilers and pressures vessels and UL in safety). Such standards require technically competent review, rather than simple public acceptance, and the circulation list is primarily to organizations and government bodies where such expertise is known to exist. The final draft is submitted to ANSI's Board of Standards Review for acceptance as an American National Standard. It is a painstaking and thorough process.

That brings us to the third method, which is the American National Standards Institute committee structure. Basically, the expertise is gathered from as wide a spread of interests as possible to form the committee which will develop the document. Once drafted, the standard receives public review and is then sent through the ANSI system for approval.

As an example of how one organization circulates its standards for approval, see App. C.

## 11.6.2 Other Conditions

*11.6.2.1 Unanimous.* A unanimous agreement is one where everybody agrees with what the standard says, the way it is written, and the belief that it will achieve the objective for which it was written. Such a condition is considered a standards writer's utopia and very rarely occurs. Generally speaking, there probably never has been a totally unanimous document.

*11.6.2.2 Majority.* A majority agreement is, in a very elementary sense, one more "aye" than "nay." It is possible to have a majority agreement among a minority faction. Consider, for instance, how a very few people, voting for a public tax measure, can bind an entire populace to paying an increased tax.

*11.6.2.3 Minority.* Minority is, of course, contrary to majority. However, there can be a minority rule as well as minority agreement.

In countries that do not have a democratic form of government, such control can center in a single individual, provided that individual has sufficient power to either motivate or coerce the majority. This is not, however, the way standards are normally created, although minority influence (as distinct from the minority control) can exert considerable leverage in any endeavor.

*11.6.2.4 Consensus.* Obviously, there can be many conditions and shades of agreement on any document. However, for most voluntary standards it is consensus approval that either sets the standard on its way, or shoots it down.

For example, assume you are a member of a committee of a dozen persons and they have worked long and hard (see paragraph 11.1.1) to bring a standard to the approval level. The committee agrees not to leave the meeting room until the final draft is approved. Since there has been some disagreement with various parts throughout the formulation process, it is not possible to get unanimous approval. Although majority approval would be fairly simple, it is contrary to the basic philosophy of standards organizations to accept a standard as an approved document if there are unresolved objections. What the committee is after is a general agreement among all the members that the document is a good, not necessarily a perfect, expression of the intended control and can be reasonably expected to function satisfactorily in avoiding or solving the problem toward which it was directed. When such agreement is reached, the committee can consider its efforts complete.

(*Note*: Recall from paragraph 11.6 that the committee itself cannot approve and issue the standard. Committee approval is only a first step toward final approval. Although I have emphasized "consensus" as the prime approval method for a standard, there are other concepts of what constitiutes approval. Some organizations feel a three-quarter affirmative response is adequate; at least one accepts "not more than 33% disapproval" as being adequate. It is understandable that for large organizations whose documents must be replied to through the mails, it would take virtually forever to get anywhere near unanimous agreement and something more pragmatic has to be accepted. The important fact is that a broad spectrum of users has been notified and given the opportunity to make their views known.)

## 11.7 Revision

It would be nice if the committee members described in paragraph 11.1.1 could lean back with the feeling that what they had created was now firmly rooted in concrete and they could take the day off. In a few cases this might be true, but experienced standards people know it is rarely so. They know a standard that is not changed and updated at frequent intervals becomes an increasingly ineffective document. If sufficiently neglected, it can become a liability rather than a benefit.

### 11.7.1 When Is Revision Necessary?

The need for revision to a standard may arise in many ways. Technology is never static; new methods, new materials, new philosophies, new social orders, ad infinitum, may demand that a standard be updated or withdrawn. Nothing stands still for very long and a period of five years is considered a maximum interval for rework of a standard. That is not to say that the committee goes off duty for five years and then comes back to the stuffy, smoke-filled amphitheater of paragraph 11.1.1 to have another go at it. The need for a change to any document may arise at any time, and may have to be incorporated into the standard without delay. In some cases the change can be effected by a "change notice" pending a general revision of the standard.

## 11.7.2 Revision Doesn't Mean the Standard Failed

Far from change being a disappointment to these dedicated individuals who have labored to produce a consensus standard, the fact that a standard is a "living" entity adds to the feeling of involvement that first brought them together. There is a fascination in dealing with documents which will work for public good by influencing or controlling some phase of human activity.

## 11.7.3 Personal Satisfaction of Committee People

It would be easy for the uninitiated person to assume that this control factor is the reason why committee people continue to make what are often personal sacrifices for the sake of standards. They might say the work imparts a feeling of importance or power. There is no justifiable basis for such thoughts.

## 12. STANDARDS ORGANIZATIONS

In paragraph 7.2.1, I said there were several hundred organizations in the United States involved with standards. I assume it to be true, but I have not tried to count these organizations. The majority are trade associations and other specialized groups. In the following pages we will learn about some of the major organizations. (See also App. D.)

(*Note*: It is not the intent of this book to describe more than a representative selection of organizations. The information provided in the text in discussions of organizations can be supplemented by the information sources listed in App. A. In that way the student can at any point make his own personal contact with any of these organizations to secure further information. The objective in providing information about these organizations is to allow a broad view of what goes on in the standards-making field.)

### 12.1 The American National Standards Institute (ANSI)

While it may be logical to infer that this organization has some sort of federal authority, it is a purely private corporation licensed under New York State law. Its purpose (very condensed) may be stated as follows: (1) to serve as the national coordinating institute for voluntary standardization and certification activities of the United States; (2) to further the voluntary standards movement as a means of advancing the national economy; (3) to insure that the interests of the public have appropriate protection and participation; (4) to provide the means of determining standards and certification programs; (5) to establish, promulgate, and administer procedures and criteria for recognition and approval of standards as American National Standards; (6) to establish procedures for recognition and accreditation of certification programs; (7) to cooperate with government agencies in achieving compatibility between government standards and voluntary standards of industry; (8) to promote knowledge and use of American national standards and accreditations; (9) to represent the interests of the United States in international, nontreaty standardization and accreditation programs; (10) to serve as a clearing house for information on standards and certification in the United States and abroad.

The purposes enumerated above cover an exceedingly broad field, but they have been formulated over an extended period of time during which the value and use of standards have become more evident. These purposes are all established with the welfare of the United States and the private individual being of paramount importance. The types of standards handled by ANSI are exceedingly varied. Anyone seeking information about standards in a particular field should contact this organization for more information. (See Apps. A and D for description of available information and complete address.)

## 12.1.1 Councils, Boards, and Committees

The American National Standards Institute operates under an extensive system of councils, boards, and committees. Some of the councils and their functions are described below.

*12.1.1.1 The Organizational Member Council.* Activities of the Organizational Member Council include such responsibilities as: (1) cooperating with the Executive Standards Council in identifying needs for new standards, or the revision of existing standards; (2) serving as a channel of communication from the organizational members to the Board of Directors; (3) submitting each year a slate of candidates to the Nominating Committee of the Board of Directors.

*12.1.1.2 Company Member Council.* The Company Member Council is composed of one member from each company which is a member of the American National Standards Institute. Its functions include: (1) advising the Board of Directors and the councils and committees, on behalf of commerce and industry in matters of policy, procedure, and planning; (2) promoting the interests of the Institute in its acceptance by commerce and industry as the source of approval for standards; (3) assisting the Board of Directors to obtain an adequate and widely representative body of members, and the needed financial support; (4) determining the needs of commerce and industry for standards and stimulating action by the institute to bring about initiation of new standards development activities; (5) advising and cooperating with the International Standards Council in its mission; (6) providing a forum for the exchange of experience which could lead to improved industrial products through standardization; (7) conducting studies and surveys to improve industrial products and practices through the proper use of standards for the benefit of science, technology, and industry; (8) assisting in identifying areas where certification programs are needed; (9) submitting each year to the Nominating Committee of the Board of Directors a slate of candidates for membership on the board to represent company members.

*12.1.1.3 Executive Standards Council.* The Executive Standards Council's duties include: (1) being alert to the need for new standards or the reexamination of existing standards as the result of changed conditions; (2) developing methods of evaluation of standards as American national standards; (3) in industry standards projects, identifying competent organizations; (4) defining the scope of proposed standards and assigning administrative responsibility; (5) stimulating and expediting work on standards projects; (6) being aware of areas affected by standards and assuring that their views are heard; (7) coordinating with the International Standards Council to assure American participation in international standards; (8) providing a channel through which any relevant member or interest may request a review of any American national standard.

*12.1.1.4 Consumer Council.* As the name suggests, the Consumer Council is mainly concerned with seeing that the consumer gets a fair deal. The council's responsibilities include: (1) providing the Board of Directors with guidance on behalf of consumers on matters of policy, procedure, and planning in support of institute objectives; (2) conducting surveys and studies to identify consumer needs for standardization of consumer goods and services; (3) making recommendations for development of standards or certification programs important to the advancement of consumer interests; (4) serving as a contact between the institute and the general public, government, and industry in the areas of standards and certification for consumer goods and services; (5) promoting the education of the consumer so that an awareness may be developed regarding the objectives of the institute relative to consumer interests; (6) assisting in achieving effective consumer representation in standards development activities; (7) submitting each year to the Board of Directors the name of one candidate to represent the consumer council on the board; (8) providing a channel through which any member of the institute may petition for a review of any American national standard.

*12.1.1.5 International Standards Council.* The International Standards Council has been given the responsibility of setting technical and administrative policies for activities involving the International Electrotechnical Commission (IEC), International Organization for Standardization (ISO), Pacific Area Standards Council (PASC), and other international standardization organizations which may be pertinent (see Apps. A and D for complete addresses of all of these organizations). Its functions include: (1) advising the Board of Directors concerning membership in international standards organizations, and the basic policy for participation in such activities (the International Standards Council acts as liaison with the Certification Committee); (2) establishing national coordinating committees to handle United States participation in the activities of international standards organizations, and approving the rules and procedures by which these committees are organized and operated; (3) reporting annually to the Board of Directors on United States participation in international standards activities.

(*Note*: Besides the councils just described, there are, among others, a Board of Standards Review and a Certification Committee.)

*12.1.1.6 Board of Standards Review.* The Board of Standards Review is the final approval authority for standards submitted for approval as American national standards. Its duties include: (1) implementing procedures for the approval and withdrawal of standards submitted as American national standards; (2) determining whether standards submitted as American national standards meet the requirements of the institute; (3) adjudicating questions and conflicts that may arise in approval procedures; (4) being watchful of the interests of those who may be affected by a particular standard so that their views will be given full and adequate consideration.

*12.1.1.7 Certification Committee.* The Certification Committee administers the authorized certification activities of the institute. Its responsibilities include: (1) advising the Board of Directors and overseeing all national activities in the field of certification; (2) fostering the development of certification programs by others in response to a demonstrated need; (3) collaborating with the government and private organizations in the development of criteria and systems for accrediting certification programs; (4) advising the International Standards Council on participation by the United States in international certification activities concerned with civilian safety, trade, and commerce.

*12.1.1.8 Other Activities.* The foregoing paragraphs provide a review of the main activities of the American National Standards Institute and its functions relative to American national standards and certification. In addition, there is a requirement that the institute shall "develop principles for the formulation of standards and the operation of certification programs, but shall not, itself, write standards."

(*Note*: ANSI, through its accredited organization method, can certify standards produced by other organizations as "American national standards.")

12.1.2 Reason for Growth

It should be understood that ANSI did not appear on the American scene total and complete as it is at the present time. It had its beginning in 1919 when the American Engineering Standards Committee (AESC) was formed by ASME, ASTM, IEEE, ASCE, and AIMME (see App. D). This was followed a year later by the addition of National Bureau of Standards (NBS) and the Army and Navy departments. In

1928, the committee was restructured and became ASA (American Standards Association). It was also opened to membership of trade associations, corporations, and other bodies, including government bureaus. During 1966-1969, the committee was again reorganized and became the American National Standards Institute. This was the beginning of emphasis of the coordinating role of the organization and de-emphasizing its part in standards creation.

The institute has done a tremendous job in furthering voluntary standards in the United States and representing U.S. interests in international standards, and while primarily concerned with civilian needs has also benefited the national interest. The ANSI organization is now being considered for further restructuring to enable it to more effectively represent U.S. interests in nongovernmental international standards organizations. It will also continue to further the development of internal standards that qualify for worldwide acceptance.

The reasons for this restructuring are based on the new expressed objective of government bodies to work with and through voluntary standards. The International Standards Council (ISC) (paragraph 12.1.1.5), would operate under an enlarged charter dealing with international standards activities, and membership in the council would include just about every interest in the United States, including representatives of local, state, and federal governments, the consumer, small business, and labor, and colleges and universities as relevant.

Again, the student can get a glimpse here of the extent and complexity of standards activities in the United States. The student should contact ANSI for further information.

## 12.2   The American Society for Testing and Materials (ASTM)

The American Society for Testing and Materials (ASTM) is quite old, having had its beginning before the turn of the century (1898) as the American Section of the International Association of Testing Materials (IATM), which was founded in Germany in 1884. In 1902 it became the American Society for Testing and Materials, but remained affiliated with IATM until the mid-1920s when the parent organization folded. Its first standard was a specification for steel rails, published in 1902. Originally the society was concerned mainly with materials, but rapidly enlarged its scope to "the development of standards and characteristics and performance of materials, products, systems, and services; and the promotion of related knowledge." In 1971 a charter revision added systems and services. There are somewhere around 24,000 individuals currently involved with committee work on ASTM standards. Since many of these people serve on more than one committee, the term "units of participation" is used to cover such involvement. At the end of 1981 there were 80,000 such units. ASTM is in effect a management system for the development of voluntary consensus standards. Since ASTM is the only member of the standards-writing community whose prime purpose is the writing of standards, its productivity is probably equal to that of all the others combined. (Recall that the American National Standards Institute performs a coordinating function for all the standards-writing organizations, but it does not itself write standards.)

### 12.2.1 Membership

Membership in ASTM is open to individuals or organizations. There are no professional requirements, since ASTM is not a professional society. Membership costs are paid either by the individual or the organization for which the individual works, but membership is not necessary in order to work on a committee. Membership does, however, provide voting privileges.

### 12.2.2 Types of ASTM Standards

There are five types of ASTM standards: (1) Standard definitions, (2) standard practices, (3) standard test methods, (4) standard classifications, and (5) standard specifications.

(*Note*: Please recall that at the beginning of this text I considered standards to be a category of control documents; it is possible, therefore, to have "standard specifications" and avoid any argument as to whether a document is a standard or a specification.*

*12.2.2.1 Gearing Up for a Standard.* When ASTM receives a request for a new standard in a field outside the scope of existing technical committees, a new committee is formed. First, a series of conferences is held in which every person or organization that can reasonably be identified as being interested in the subject is invited to participate. The new committee is formed only after recommendation by a consensus of these persons at an organizational conference. When the committee's title and scope are approved by the Board of Directors, the new committee is in business. While committee membership is open to any concerned person, the only restriction is that producer interests cannot constitute a majority of the membership; thus they cannot dominate decisions.

*12.2.2.2 The Technical Committee.* The "blood, sweat and tears" of committee work was protrayed in paragraph 11.1.1. There is not much difference between committees. A group of people is given an objective of providing a document to serve a public need, and they go to work.

---

*See my article Standards and specifications: Engineering's Siamese twins, published in *Engineering Graphics* (now *Design, Drafting and Reprographics*) in February 1975.

*12. 2. 2. 3 Progression.* In ASTM, a standard moves from the originating task force to a subcommittee, then to the main committee, and finally to the society as a whole (by which is meant that all members can vote to approve or disapprove the standard). At each step the action is formalized by letter ballot. Every effort is made to resolve negative votes before proceeding to the next step.

*12. 2. 2. 4 Approval.* Once the society letter ballot is completed acceptably and the ASTM Committee on Standards has been satisfied that the procedural requirements have been met, the standard is considered approved. (See paragraph 11.6.)

*12. 2. 2. 5 Publication.* When completed and approved, all new standards are published in the *Annual Book of ASTM Standards*. Once published, the standard remains under the jurisdiction of the Technical Committee, and the committee is responsible for keeping it current with changing technology.

The *Annual Book of ASTM Standards* at this time consists of 48 books or "parts," with part 48 being a combined index. The standards in each book are assembled in alphanumeric sequence by their ASTM designation numbers, except volumes 31 and 47, which are assembled by subject matter. Parts 25, 26, and 29 are assembled first by committee identification, then alphanumerically. For user convenience, each part has two tables of contents, one being a list of standards in numerical sequence in the ASTM designation, and the other a list of standards classified by subject. At the end of each part is a subject index. Practically all libraries have sets of ASTM standards, so it is not difficult for students to consult them for whatever they need.

(*Note*: In past years many ASTM standards were submitted to ANSI for accreditation as American national standards. The reader will find such standards identified jointly as ANSI/ASTM standards. Many standards of other organizations have been similarly identified. Publication rights remain with the developer and ANSI acts in a marketing role. (See paragraphs 11.6.1 and 12.1.1.8.) As this is written, some changes are taking place and future methods may differ.

*12. 2. 2. 6 Revision.* At least once every five years the committee must review the standard and recommend it for reapproval, with or without revision, or recommend its withdrawal.

(*Note*: The foregoing section provides a look at standards and their development in ASTM. The student should seek further information from ASTM. See Apps. A and D for address.)

12. 2.3 The Forward Look

In an effort to keep up with the accelerating pace of standardization, the ASTM Board of Directors each year adopts a long-range plan for the society covering the next five years. The objectives are (1) to keep the total membership and the public informed about how the directors and staff view the future; (2) to provide an overview of the challenges and opportunities facing the society; and (3) to bring a wider interest and expertise to bear on the evaluation of future developments.

*12. 2. 3. 1 Maintaining Efficiency of Committees.* No organization and no industry can ignore the reality of standards and their effect on the future economic welfare of the country. Within ASTM the Board Committee on Technical Committee Activities (TCA) is charged with maintaining committee efficiency and productivity.

*12. 2. 3. 2 Continuous Need for Documents.* It is expected that there will always be new waves of social, industrial, and political issues coming to public consciousness, and it is the responsibility of the voluntary standards organizations to provide the documents for an orderly development.

(*Note:* There have been lawmakers who claim to have seen isolated cases where a standard has resulted in some discrimination against a small business or a consumer, and have proposed laws to allow federal control of voluntary standards. This step would be a disservice to the country, although certainly it is obligatory for the government to be aware of, and be involved with, standards, particularly those affecting international trade.)

## 12.2.4 New Fields

New fields are opening in energy consumption and generation, space heating (including solar), transportation, and industrial processing.

## 12.2.5 Medical and Health Care Services Standards

Medical and health-care services need extensive attention, and standards must be available for medical materials, devices, components, systems, and procedures. Standards are also needed in the field of nutrition. The fact is that whenever a new field of human endeavor opens, there must be standards to control and guide its progress and record its history for those who follow.

## 12.2.6 Consensus Standards

With increasing federal, state, and local attention to standards, the concept of consensus standards becomes ever more important, and special attention must be paid to maintaining an equitable balance between different interests. In this respect, standards become something of an umpire or referee in seeing that no one gets special treatment or advantage.

*12.2.6.1 Federal Controls.* Surely the student is aware of the vast number of federal controls associated with environmental conditions, worker safety, and protection of the consumer. When the government creates mandatory standards it is generally because there are no voluntary standards, or none sufficiently exact in nature, to cover the need.

## 12.2.7 Consumerism

In the field of consumerism, it is also essential to maintain a balance between interests. Important as the consumer is, without the manufacturer there would be no jobs and no products for the consumer to consume. There have been instances where small businesses have been forced to suspend operations because they could not meet the restrictions placed on them by federal bureaus in the interest of the consumer. It follows then that creating standards that will equally protect all interests is an exceedingly important, and difficult, job.

## 12.2.8 The Academic Community

ASTM, as well as other standards organizations, has long been aware that education in standards is practically nonexistent. The society has conducted seminars in an endeavor to make instructors more aware of standards so that they can pass the information on to students. This undertaking is commendable and, if it is supported by informative texts there should be an awakening of academic interest.

(*Note*: The student should understand that ASTM standards are identified by the organization as "voluntary consensus standards." This means their use is entirely voluntary, and the existence of an ASTM standard does not preclude anyone from "doing his own thing." Now, in spite of the fact this is a pretty standard statement for all the voluntary standards organizations, the fact is nobody can function without standards, so when something is being made, marketed, purchased, there is a search for applicable standards. Such standards, once found, are then used as reference. The end result is that "voluntary consensus standards" are a great deal less voluntary than the concept on which they are based. This is true of any standard found applicable to a given situation.)

## 12.2.9 International Standards

The activities of ASTM (as an organization) in relation to the development and adoption of ISO and other international standards is minimal. Some ASTM committees provide the secretariats for ISO technical committees; others support working groups. ANSI is relied upon to represent the interests of the U.S. standards system.

(*Note*: While the foregoing is far from an exhaustive coverage of ASTM, it is necessary to provide coverage for other organizations as well. Keep in mind that the objective of this text is to give the student a general overview of the field of standardization, rather than trying to provide in-depth coverage for every organization.)

## 12.3  American Society of Mechanical Engineers (ASME)

Like many other technical societies, the American Society of Mechanical Engineers (ASME) began as an educational and technical society of individual members (1882). Although it still carries on these functions, ASME activities now encompass vastly greater areas than its originators dreamed (see Apps. A and D for complete address).

### 12.3.1  Goals

The principal aims and objectives of ASME, briefly stated, are: (1) to provide continuing education to mechanical engineers, the industries they serve, and society in general, through the development and dissemination of technical information; (2) to develop mechanical standards, codes, safety procedures, and operating principles; (3) to increase the personal and professional development of practicing and student members; and (4) to aid members of the engineering profession in maintaining a high level of ethical conduct.

### 12.3.2  Membership

There are over 81,000 members, of which 12,000 are student members.

### 12.3.3  Scope

There are very few fields which are not touched by mechanical engineering. Subjects range from solid waste processing to space exploration.

### 12.3.4  Publications

ASME conducts one of the world's largest technical society publishing operations, involving a vast spectrum of engineering experience and research.

*12.3.4.1  Quarterlies.*  Transactions of ASME are published in eleven quarterlies, which include the *Journal of Applied Mechanics, Journal of Engineering and Industry,* and the *Journal for Pressure Vessel Technology.* The monthly magazine, *Mechanical Engineering,* and various other standards, codes, and handbooks, along with applied mechanical reviews, carry the message of mechanical engineering.

*12.3.4.2  Codes and Standards.*  Since its inception, the society has taken the lead in establishing codes and standards which have been the backbone of the nation's industrial strength. They have also been used internationally. Presently, over 14,000 members are engaged in the work of creating new standards and revising and updating old codes and standards.

*12.3.4.3 Research.* Although the society does not maintain research laboratories, it consistently sponsors research projects. First, a special advisory committee is formed to initiate a research project, after which an established research organization is contacted to do the actual tests, experiments, or whatever is required.

## 12.3.5 Education

ASME is directly involved with the Engineers Council for Professional Development (ECPD) and the American Society for Engineering Education (ASEE). ECPD is responsible for the national accreditation of engineering curriculum. ASEE (see App. D) is primarily involved with the content of the curriculum, the methods of engineering teaching, and the continuing education of the practicing engineer.

## 12.3.6 Basic Engineering Department

The Basic Engineering Department deals with applied mechanics, fluids engineering, ocean engineering, and technology and society, among some fifteen disciplines.

## 12.3.7 Industrial Department

The society maintains seven industrial departments, which include such diverse fields as aerospace, petroleum, and rail transportation.

## 12.3.8 Power Department

The Power Department has eight divisions, among which are air pollution control, energetics, and solar energy.

## 12.3.9 Technical Standards

Some fields of technical standards activity include safety standards, boiler and pressure vessel codes, performance test codes, and industrial standardization. The major activities are with boiler and pressure vessels and performance test codes for such things as turbines, steam engines, incinerators, and nuclear systems.

## 12.3.10 Intersociety Activities

ASME interfaces with a multitude of other engineering disciplines, both nationally and internationally. A few of these are: Council of Engineering and Scientific Society Executives (CESSE), Federation of Materials Society (FMS), National Government Affairs Liaison Committee, Pan American Federation of Engineering Societies (UPADI)

(EJC), and the World Federation of Engineering Organizations (WFEO).

(*Note*: Any student studying mechanical engineering is going to be involved with ASME. As stated previously, because of the pervasive nature of standards matters, there is a practical impossibility in trying to sharply define the limits of any single standards activity.)

## 12.4  National Electrical Manufacturers Association (NEMA)

The National Electrical Manufacturers Association (NEMA) was started about 1905 (see Apps. A and D for address). Today it is a very large nonprofit trade association of manufacturers of electrical apparatus and supplies. Its stated objective is to promote the standardization of electrical apparatus and supplies and to facilitate understanding between manufacturers and users of electrical products.

### 12.4.1  Subdivisions

The parent organization is divided into various subdivisions which have the responsibility for a category of apparatus or supplies (e.g., fire protection).

### 12.4.2  Purpose of NEMA Standards

Standards promulgated by NEMA are adopted in the public interest. They are designed to eliminate misunderstandings among those involved with electrical products, and to assist the purchaser in selecting and obtaining the proper product for his particular need. Lack of conformity to an NEMA standard does not preclude any member or nonmember from manufacturing and selling his products.

(*Note*: While this is basically correct, if a document references a product to an NEMA standard and the product does not conform, it may not be acceptable to a particular user. This is likewise true of most products covered by voluntary standards.)

### 12.4.3  Defining an NEMA Standard

A standard document of the NEMA may define a product, process, or procedure, with reference to nomenclature, composition, construction, dimensions, tolerances, safety, operating characteristics, performance, quality, rating, testing, or service. (Now that I've listed all these, what else could the NEMA standards cover, except possible the color of electricity?)

### 12.4.4  Classes of Standards

NEMA standards fall into one of two classes. (1) A standard which relates to a product commercially standardized and subject to repetitive manufacture. Such standards must be approved by at least 90 percent of the members of the involved subdivision. (2) Proposed standards for future design. Such documents may not have been originally applied to a commercial product, but embody a sound engineering approach to future product design. Their use depends on approval of at least two-thirds of the membership of the responsible subdivision. These standards are generally based on, or are compatible with, Underwriters' Laboratories (UL), ASTM, and military standards.

For further information, write NEMA.

## 12.5 Electronic Industries Association (EIA)

The student will probably wonder why so many different organizations are necessary to deal with electrical subjects. The Electronic Industries Association (EIA), like NEMA, is a nonprofit organization representing manufacturers of electronic products (see Apps. A and D for address). As a national organization, EIA confines its activities to the area of legitimate public interest objectives under the direction of the Board of Governors.

### 12.5.1 Intentions

It is intended that all activities of the EIA shall stimulate public awareness of the vital role of electronics in the national defense, space exploration, communication, education, and entertainment. It also aims to create awareness of the evolution of industrial technology, the importance of having standards (based on electronic development), and the effect of electronics on the gross national product of the United States.

### 12.5.2 EIA and Industry

The EIA intends to provide a forum within our national laws and policies where industrial representatives can discuss matters affecting the legitimate interests of their industries and the implementation of the policies of the association.

### 12.5.3 EIA and the Military

Activities of the EIA are intended to assist the Department of Defense (DOD) and the armed forces in obtaining the most advanced and reliable products and scientific development from industry through an interchange of information and ideas.

### 12.5.4 Dissemination of Technology

It is further intended to advance the growth and technological progress of the industry by providing facilities and staff assistance in the development and dissemination of technical and related information, the registration of new products, and the participation in national and industrial standardization.

## 12.5.5  Accomplishment of Objectives

EIA accompoishes its objectives through seven product-oriented divisions and two major departments, the Engineering Department and the Marketing Services Department.  As an idea of its scope, there are nearly 300 member companies which provide for about 200 committees and working groups in domestic and international standards development.

(*Note*: The standards situation changes so rapidly that any figures given today are seldom the same a week hence.)

Participation in EIA standards committees, as with all other voluntary standards organizations, is open to any qualified person who is sincerely interested in the objectives of the association.

## 12.6  Institute of Electrical and Electronic Engineers (IEEE)

The fact that there are so many organizations concerned with electrical and electronic subjects is evidence of the very great importance of this branch of technology.  The Institute of Electrical and Electronic Engineers (IEEE) intends to advance the theory and product of electrical engineering, electronics, radio, and allied branches of engineering in all related arts and sciences.

## 12.6.1  History

IEEE is the result of the merger of two earlier organizations, the Institute of Radio Engineers (IRE) and the American Institute of Electrical Engineers (AIEE).  Its activities are oriented toward creativity in the electrical and electronic fields.  As the name indicates, the emphasis is on the engineering aspects of the field.  It is the world's largest professional engineering society.

## 12.6.2  Hypothesis

The hypothesis of the institute is that engineers in a professional group can contribute most effectively in the area of basic technical subject matter, and that proposals for basic improvement in the electrical and electronic fields are a natural outcome.

## 12.6.3  Objective

It is intended that any IEEE standard should clearly answer the question "What?"  It should provide a specific basis for ratings and tests to determine performance of either specific equipment or materials.

*12.6.3.1 Scope of Standards.* IEE standards are used widely throughout the United States and are an important part of the electrical and electronic "standards package" for manufacturers and users of related equipment.

*12.6.3.2 Proposed Standards.* Not all documents issued by the IEEE are considered standards. There are secondary documents, such as guides, which are identified as proposed standards. These documents are issued as exploratory standards for the purpose of gaining practical experience in a particular field, or a better definition of a product, and they may later evolve into accepted standards.

### 12.6.4 Liaisons with Other Standards Organizations

In order to not overlap the activities of other organizations, IEEE maintains a liaison network and encourages its members to become involved with committees of other standards organizations. This is likewise true of all large standards-writing organizations.

### 12.6.5 Educational Activities

The institute also partakes of the nature of an educational unit, and provides courses in a broad variety of subjects including computer technology. The courses are usually one- to five-day seminars and are taught by people from industry, from universities, and from other national organizations. The broad objective is to allow engineers to keep abreast of developments in the electrical and electronic engineering field. Home study courses are also made available. All such courses are produced by the IEEE Educational Activities Board. For information contact the IEEE.

(*Note*: What I have covered here is barely the tip of the IEEE iceberg; this also applies to other organizations described in this text. Addresses are given in Apps. A and D for use by the student in securing additional data.)

## 12.7 International Organizations

The international standards organizations in this section are not introduced in any order of importance. All standards organizations, whether large or small, domestic or international, are important.

### 12.7.1 International Electrotechnical Commission (IEC)

The International Electrotechnical Commission (IEC) has been in existence for a long time (see Apps. A and D for complete address). It first met in 1908. It is difficult to be accurate about the number of member countries, since the membership increases continually. Many new nations have appeared in recent times, and the membership is about 50 countries at this time. These nations represent 80 percent of the world population and 95 percent of the world production and consumption of electricity.

*12.7.1.1 Objectives.* Electrical engineers were among the first to comprehend that international standards were going to be a fact of life (a very important fact), and there were meetings even before the turn of the century. The commission is intended to facilitate the coordination and unification of national electrotechnical standards. The recommendations which the commission makes are not binding on the member countries, but (as I explained in paragraph 12.4.2) even voluntary standards can have the force of mandatory standards if they are observed by a sufficient number of users, and are referenced as applicable to particular situations.

*12.7.1.2 Scope.* The work of the IEC covers almost all spheres of electrotechnology, including telecommunications and nuclear energy. The work can be divided into two parts (1) improving understanding between electrical engineers of all countries by drawing up common means of expression, nomenclature, agreement of qualities of units, symbols, abbreviations, and graphic symbols for diagrams, and (2) standardizing electrical equipment, involving the study of the problems of the electrical properties of materials used in electrical equipment, standardization of equipment characteristics, methods of test, quality, safety, and interchangeability of electrical equipment.

(*Note*: While electricity can give a shock in any language, the amount of electricity allowed to flow in lines and to activate equipment may be different in various countries. It is well for the traveler to be prepared to adjust to different voltages through the use of converters. Advance information about voltages can be very helpful. IEC standards are written in English, French, and Russian. There are 1230 publications representing more than 39,000 pages of standards, plus reports, modifications, etc., listed in the IEC catalog.)

*12.7.1.3 Administration.* The commission is administered by a council on which all the national committees are represented. This council might be likened to the United Nations. Detailed technological work is carried on by technical committees assigned to particular subjects. There are 188 technical committees and 500 working groups. About 250 draft standards are submitted per year, and over 100 meetings take place in various parts of the world.

*12.7.1.4 Liaison with National Organizations.* The commission maintains liaison with those national organizations which deal directly or indirectly with electrotechnology.

*12. 7. 1. 5  Procedures for Approval of Standards.*  A postal vote among all the national committees is necessary for approval of any standard. Documents are considered approved if no more than one-fifth of the national committees cast a negative vote.  (Apparently it may be easier to get less than one-fifth negative than to get four-fifths positive.) Two-thirds approval is about run-of-the-mill for most general types of standards, but don't forget the principle of consensus approval (see paragraph 11.6).

*12. 7. 1. 6  Recommendation for Internationalization of Standards.* Recommendations are made when approved by the national committees, and the documents are intended to serve as the basis of the national standards of the concerned countries.  Pending formal approval, a report may be published as an intermediate document.  The IEC can be contacted for further information.

12. 7. 2   International Organization for Standardization (ISO)

The International Organization for Standardization (ISO) is a specialized international agency created under the auspices of the United Nations in 1947 and aimed at worldwide agreement on international standards. Its objectives include expanding trade, improving quality, increasing productivity, and reducing costs of goods and services.  It is the only international organization in this field.  Regional bodies have been formed for reaching agreement among countries.  These include the Pan American Standards Commission (COPANT) in Latin America, the European Committee for Electrotechnical Standardization (CENELEC) in the Common Market countries, and the Pacific Areas Standards Congress (PASC) in the Southeast Pacific Basin. (See App. D.)

*12. 7. 2. 1  Contributions.*  ISO issues more than 500 new and revised standards each year, representing a contribution to the collective wealth of technical and scientific experience throughout the world.

*12. 7. 2. 2  Membership.*  There are currently 82 member countries, including the Peoples Republic of China and the Soviet Union.  These countries represent 95 percent of the world's industrial production. One "national" body from each country is allowed membership.  Most such bodies are governmental institutes or organizations incorporated by public law.  For new countries which do not have national-stature standards bodies, there is a special membership called correspondent membership.  Usually, a correspondent member becomes a regular member within a few years (as standards development matures in that country).

*12. 7. 2. 3  History.*  Until a relatively few years ago, international standardization was almost solely the concern of industries, both as producers and consumers of goods and services. The first ISO committees dealt with screw threads (which are still being internationalized), bolts, nuts, and other hardware.

12.7.2.4 *Present Scope.* Today the scope of ISO has broadened to include just about everything that passes between nations, and encompasses virtually all fields of human endeavor. There are approximately 300 main technical committees and around 1700 working parties (groups). While industry is still heavily involved, interest in international standards today also comes from government agencies, the scientific community, and other international bodies and consumer agencies.

12.7.2.5 *Participants.* Would you believe 100,000 experts from all over the world are involved in ISO technical work? Of this work, the main part is done by correspondence, and each year the committee secretariats distribute for comment more than 10,000 working documents.

12.7.2.6 *Technical Meetings.* There are about 800 ISO technical meetings held yearly in some 40 countries and attended by 20,000 delegates. Meetings last from two or three days to two weeks. The combined efforts of these committees has resulted in more than 3000 published and revised standards. It is estimated the cost of this involvement is equal to several hundred million Swiss francs. Since the value of the franc varies daily, it is not possible to state the true dollar value, but it is quite a few million dollars.

(*Note:* Remember that this gigantic effort is based on voluntary efforts of thousands of people who believe in the value of standards to the world in which they live. This is not to say there are no personal interests involved, since these people often represent the economic interests of their companies. However, very few of these people receive any money for interacting with standards bodies.)

12.7.2.7 *Acceptance of an International Standard.* An international standard is the result of an agreement between ISO member bodies, with a 75 percent favorable vote required to accept a draft proposal before it can be sent to the ISO council for acceptance as an international standard.

(*Note:* The student might wonder by this time why these thousands of hardworking individuals continue to labor not only willingly, but enthusiastically, for good standards. I refer you back to the concluding lines of paragraph 11.1.1.)

12.7.2.8 *Adoption of International Standards.* Some international standards are adopted outright as national standards (some countries are even relying totally on international standards), others are incorporated into, or coordinated with, national standards. However it works out, the beneficial results of international cooperation on standards is a strong force for peaceful coexistence. It is, in its own way, a united nations assembly.

*12. 7. 2. 9   Distribution of Workload.*   The actual technical work on international standards is divided among technical committees, sub-committees, and working groups. Secretariats (convenors or chairs of technical committees) are distributed among the ISO member bodies, and the work of these bodies is coordinated by a central secretariat.

*12. 7. 2. 10   The Board of Directors.*   This board is an elected council that meets once a year. It is made up of representatives from 18 member bodies.

*12. 7. 2. 11   The General Assembly.*   This group of delegates, nominated by all ISO member bodies, is the principal policy-making body. The GA meets every three years.

(*Note*:   By this time, it might be thought I have given a pretty fair coverage of standards. The truth is that there is still a great deal to say. Even when this text is marked "finis", it will have only lightly skimmed over the total worldwide field, but it is hoped that it will have stimulated you to seek more information from ISO and other organizations listed in Apps. A and D.)

## 12.8   The National Bureau of Standards (NBS)

While the National Bureau of Standards (NBS) is not in the same "family" as voluntary standards organizations, it would surely never do to leave this bureau out of any discourse on standards.

### 12.8.1   History

In 1901, in a world where any type of standard was apt to be a purely local control, where a pound of butter might have several different weights at various stores in the same block, and where most of the scales and measuring devices were still sent abroad for calibration, the United States government created the National Bureau of Standards to provide the basis for the orderly conduct of industry and commerce. (Note that word, orderly, it is one of the prime benefits of standards.)

### 12.8.2   Proposed Function

The legislation establishing the bureau described its intended function as: ". . . shall consist of the custody of standards, the comparison of the standards used in scientific investigation, engineering, manufacturing, commerce, and the educational institutions . . . determination of physical constants and properties of materials . . . measurements which are not to be obtained of sufficient accuracy elsewhere." This clearly established the bureau as the custodian of physical standards, which it remains today.

12.8.3   Expanded Scope

However, as the years passed, the scope of the bureau expanded until today it is also deeply involved in the performance of scientific research, test methods, and standards-writing in such areas as energy conservation, fire safety, computer applications, environmental protection, materials utilization, and consumer product safety and performance. (See also UL, paragraph 12.10.)

12.8.4   Role as Catalyst for Advanced Technology

For three-quarters of a century, the NBS has served as a catalyst for the advanced technology needed by American science and industry.

12.8.5   Role as Coordinator of Product Standards

The bureau also coordinates the development of voluntary product standards, if such standards do not already exist in the private sector. NBS has collaborated on more than 100 nationally recognized requirements for particular products (ranging from glass bottles to toys) that the industries involved have agreed to follow. These standards play an important role in the marketplace by insuring better and safer product performance.

12.8.6   Interaction with Consumers

Consumers are brought directly into the standards-making process through a nationwide network of consumer sounding boards. These boards are sponsored by NBS, ASTM, NFPA, and UL/AHAM (see App. D). ANSI coordinates the activities of these boards, which provide standards organizations with direct grassroots opinions of those who use consumer products.

12.8.7   Research Activities

The bureau performs work for many government and private organizations upon request by conducting investigations of phenomena which the requesting area cannot readily conduct themselves.

12.8.8   Dissemination of Information

Since it was formed, the bureau has been providing information on countless subjects to equally countless requestors. During a single year, NBS received 31,000 letters and 13,900 telephone calls on subjects ranging from energy conservation to standard colors. (It is amazing, for instance, how many colors are "red.") Recently, the NBS has been deluged with requests for information on metrics. This pressure resulted in the establishment of a Metric Speakers Bureau of more than 170 persons, with at least one available in every state.

## 12.8.9 Reorganization of NBS

In 1978, the bureau underwent major reorganization to enhance the effectiveness of the work it performs. The new organization was along major functional lines and three institutes were abolished. These were the Institute for Materials Research, the Institute for Applied Technology, and the Institute for Basic Standards. In the process, two new groups were formed: The National Measurements Laboratory (NML), and the National Engineering Laboratory (NEL). The NML is basically a measurement methods-standards-data organization that performs physical science research. It also provides advisory and research service for other government agencies. The NEL performs measurement-related research.

While it was not directly intended that either of the above groups would be involved with standards-making activities, such activity was inevitable from the nature of the conducted research. Since research is a function that reaches out toward the new and unknown, the results of the activity must be written, with various standards as the natural outgrowth.

An area not substantially changed by the reorganization is the Institute for Computer Science and Technology. At the time of reorganization, the government's automatic data processing (ADP) expenditures were about ten billion dollars and involved over 10,000 computers and 150,000 people, and it is in this field that the Institute for Computer Science and Technology performs its work or research and development. The institute develops federal ADP standards and guidelines.

## 12.8.10 Publications

Publications from NBS are important sources of information for consumers, businesses, government organizations, universities, and professional societies. More than 40,000 pages of research have been published in a single year. Beyond a doubt, the National Bureau of Standards is the preeminent disseminator of information concerning standards. Possibly I should qualify that to be "physical standards and measurement" since some of the other standards groups are specialists in particular fields. NBS welcomes letters and requests for information.

---

*For the student wishing to understand more about the economics of standardization, a report written and published by the NBS (stock number PB 81-120362) can be ordered from the National Technical Information Service (NTIS). (See App. D for address.)

## 12.9 Instrument Society of America (ISA)

The Instrument Society of America (ISA) was founded in 1945 and has grown to quite a large association with about 150 sections in various geographical areas of the United States. (See Apps. A and D for complete address.) There are also members in other countries. The society's structure and activities very closely parallel those of the Standards Engineering Society (SES, see paragraph 13).

### 12.9.1 Objectives

The ISA is a nonprofit scientific, technical, and educational organization dedicated to advancing the knowledge and practices related to the theory, design, manufacture, and use of instruments for measurement and automatic control in science and industry.

### 12.9.2 Scope

The society is concerned with the application of all aspects of instrumentation to industrial, laboratory, biophysical, marine, and space environments.

### 12.9.3 Functions

Technical standards are developed by technical committees composed of volunteers from government and industry, and the standards are publicly reviewed to insure a consensus view (see App. C). The standards encompass subjects such as thermocouples, flowmeters, flow plan symbols, hazardous environments, dynamic testing, and various other types of measuring equipment. The society cooperates with the U.S. Department of Health, Education, and Welfare to develop curriculum guidelines for instrumentation instruction.

(Note: The reader will have observed the nonprofit status of voluntary standards organizations. It is difficult to see how a business could be established to write standards for profit, because the element of mandatory use would be lacking. Very shortly one would expect such a venture to become "nonprofitable". This does not, of course, apply equally to standards-marketing organizations.)

## 12.10 Underwriters' Laboratories, Inc. (UL)

Practically everyone knows the UL label. Underwriters' Laboratories, Inc. (see Apps. A and D for complete address), is a nonprofit organization without capital stock. It is chartered to establish, maintain, and operate laboratories for the examination and testing of devices, systems, and materials. Dating from 1894, the enterprise has operated for service, not profit.

(*Note*: The student should be conscious of the fact that all organizations, whether "for profit" or "not for profit," tend to be self-perpetuating, and those where salaries are paid to various personnel do have "income incentive." This is in no way intended to be a criticism, but simply to inform the student that self-interest is not necessarily absent in nonprofit organizations. Essentially, this is a good thing; if there were no self-interest, there would be a constant turnover of "core" people and a weakening of the whole effort.)

### 12.10.1  Objectives of UL

*12.10.1.1  Laboratory.*  The formally stated objectives of UL are extremely detailed. They are to conduct scientific investigations, experiments, and tests, on materials, methods, products, and systems, to determine whether hazards to life or property exist, and to publish standards and other documents that will reduce or prevent any such loss of life or property.

*12.10.1.2  Public Service.*  The objectives are further stated as (1) to contract with industry, governmental agencies, and others for the purpose of conducting such experiments, tests, and so on; (2) to report and circulate the results of such experiments and tests to insurance organizations, public safety authorities, government bodies, and other interested parties, and the general public, by the publication of lists and descriptions of such materials and products; and (3) to provide for the attachment of markings or labels, or to issue certificates or such other means of identification as may be appropriate.

### 12.10.2  Fields Covered

Fields covered by UL activities include (1) electrical appliances, devices, systems, and materials; (2) accident-prevention products; (3) products which may have inherent hazards; (4) burglary-prevention systems and equipment, and other signaling equipment; (5) fire-resistant building materials; (6) fire-extinguishing devices and systems; (7) gas and oil equipment; (8) electrical and other equipment for use in hazardous atmospheres; (9) chemical products and processes; (10) medical, dental, x-ray, marine products, and other equipment.

### 12.10.3  Engineering Services

In 1981, UL engineering departments associated with the foregoing subjects handled 54,429 new work assignments to conduct investigations (for safety) of products, materials, and systems.

### 12.10.4  Inspection Services

As many as 295,690 factory follow-up inspections to various domestic companies were conducted by UL in 1981 to verify the means employed by manufacturers to establish that listed products remain in conformance. These visits are not scheduled, and the UL representatives inspect production-line products for conformity with UL requirements. Several billion labels bearing UL's registered mark were used by manufacturers to identify complying products. In addition, 11,240 investigations were conducted in 1981 by UL for foreign manufacturers on products intended to be imported to the United States. Underwriters' Laboratories maintains 59 inspection centers in 54 foreign countries covering 8800 factories.

### 12.10.5 Standards

UL publishes more than 450 standards for safety of materials, devices, construction, and methods. These standards are issued directly by UL and may also be obtained from the American National Standards Institute (ANSI).

*12.10.5.1  Standards Progression in UL.*  UL submits a standard to ANSI as an accredited sponsor using ANSI's canvass method, or it may use the accredited organization method of ANSI to develop a standard. Prior to submitting a proposed standard to ANSI under the accredited organization method, UL proceeds through a series of steps in developing a standard as follows:

1. *Announcement of the proposed standard.*  When a project is initiated by UL, announcement of its intent is sent to a broad selection of organizations. The American National Standards Institute is asked to assist in preparing the list, which includes
   a. UL subscribers
   b. Known representatives of nonsubscriber manufacturers and national interests that are concerned with the subject of the proposed standard
   c. Various UL councils
   d. Identified industrial and commercial groups
   e. Representatives of ANSI
2. *Technical advisory groups.*  When an initial draft of a proposed standard is completed by UL and an industry advisory committee (UL-IAC), it is submitted to the UL-Technical Advisory Group (UL-TAG) for review. This first review is intended to determine that the provisions of the standard provide proper guarding against the hazards inherent in the product or in its subsequent use, and that the product can be installed and used in accordance with nationally recognized codes.
3. *Product safety standards committee.*  A draft standard approved by the UL-TAG is submitted to the UL-Product Safety Standards Committee (UL-PSSC) for review to establish that the draft standard meets the guidelines listed in the preceding paragraph. A summary report is prepared including a new draft which may be

modified as a result of action by the UL-PSSC or UL-TAG. When it is complete, copies of the report are sent to all those who expressed interest in the project to develop the standard.

4. *Revision.* When comments are in from this circulation, they are evaluated. If a revised draft of the proposed standard is considered necessary, it is sent out to substantially the same people as the previous circulation. If necessary, a new summary report will follow.

5. *ANSI activity.* When the activities described above are complete, ANSI is requested to publish in its widely circulated *Standards Action* that a draft of such a proposed standard is available for public comment.

(*Note:* The information presented here for UL, and that given previously for other organizations, should make it apparent that the process of creating a standard by the consensus method is not a quick one-two operation, but a lengthy, complicated, and time-consuming effort that endeavors to inform as many persons as reasonably possible. The time span for completing a standard by consensus has been held by some critics as being a defect of the system. In that regard, so is the democratic form of United States government. The alternative is what, mandated standards?)

6. *Standard Review Council.* The Standard Review Council of UL (UL-SRC) has the responsibility of deciding that the proposed standard has had a high degree of acceptance by those who have been involved and that it meets the UL procedures for development of a product safety standard under the accredited organization method of the American National Standards Institute. (See paragraph below.)

7. *ANSI Board of Standards Review.* Under the accredited organization method, when the UL Standard Review Council is satisfied that the proposed standard has met the necessary criteria, it is submitted to the ANSI Board of Standards Review (BSR) as a proposed American national standard.

(*Note:* There are 63 organizations at this time that submit standards to ANSI for publication. The major organizations (those with the most standards) include the American Society of Mechanical Engineers (ASME), the American Society for Testing and Materials (ASTM, see 12.2.2.5 Note), the Electronic Industries Association (EIA), the Institute of Electrical and Electronic Engineers (IEEE), the National Fire Protection Association (NFPA), the Society of Automotive Engineers (SAE), and UL.)

8. *Acceptance for publication.* When the ANSI Board of Standards Review approves a standard as an American national standard, the standard will be marketed by UL and may also be published by ANSI.

(*Note*: Although American National Standards Institute standards are made available to anyone who desires a copy, they
are usually not distributed free. The cost varies with the
complexity of the subject. The money received provides some
organizations with a substantial part of their operating capital.
ANSI buys and sells standards in the capacity of a broker, rather
than a publisher.)

9. *Upkeep of UL standards.* UL (as well as the other organizations)
is responsible for maintaining its own standards. This means that
revisions must be made as necessary, and the standard is cri-
tically reviewed at intervals of not more than five years.

(*Note*: Standards which are reviewed may not need to be revised;
however, they can be reissued as "reaffirmed" on such and such
date. In this way the users are informed that the standard, however old its original issue date, is still a valid document.)

10. *Public involvement of UL.* A distinguishing characteristic of UL
is its heavy involvement with outside agencies and groups in
addition to its subscribers. This involvement takes the form of
participation in technical committees that develop product installation, maintenance, and use standards. These standards have
great impact on UL because the products which UL lists and
classifies are required to be designed so that they can be installed and used in accordance with these standards.

11. *Regulatory agencies.* A second area of major involvement is with
regulatory authorities including electrical inspectors, building
officials, and fire and police groups. A large number of these
authorities are represented on UL engineering councils.

12. *National Fire Protection Association (NFPA).* The standards
organization with which UL has the most contacts is the National
Fire Protection Association (NFPA). Approximately 50 staff
people serve on 100 technical committees concerned with fire protection. Information may be obtained from NFPA (see App. D for
complete address).

13. *American National Standards Institute.* UL very strongly sup-
ports ANSI, and about 40 UL representatives serve on 69 standards committees. Under the auspices of the institute, UL has
taken an active role in the U.S. National Committee of the International Electrotechnical Commission (IEC).

14. *American Society for Testing and Materials.* At the present time,
29 UL representatives participate in 31 ASTM committees, many
of which relate to development of test methods. Some UL and
ASTM standards are nearly identical; however, UL standards have
the objective of establishing criteria for judging the acceptability
of a product for safety, while ASTM standards spell out test
methods and do not necessarily include criteria for acceptance of
a product.

15. *American Insurance Association (AIA)*.   The AIA (see App. D
    for complete address) has a number of technical committees which
    provide input in specified engineering areas for the insurance
    code-making process.   UL has representatives on the AIA Engin-
    eering and Safety Committee.
16. *Building codes.*   Underwriters' Laboratories interacts with such
    other code-making bodies as the International Conference of
    Building Officials (ICBO), the Building Officials and Code Ad-
    ministrators International, Inc.   (BOCA), the Southern Building
    Code Congress (SBCC), and the National Conference of States on
    Building Codes and Standards (NCSBCS).

    (*Note*: The student might as well become resigned to living in an
    alphabetical wonderland, since the unhandy and tedious use of
    full organizational names would make any document virtually un-
    readable.   On the other hand, use of abbreviations and short
    forms without adequate identification is a sin standards people
    should not indulge in.   See App. D.)

17. *Federal government.*   For more than 40 years, the UL staff mem-
    bers have had responsibility for liaison with the federal govern-
    ment.   UL standards are referenced in federal specification, and
    in some cases UL has modified its standards to suit federal re-
    quirements (but never by relaxing the requirements for safety).
    Federal agencies personnel are members of UL councils.

    (*Note*: This has been a somewhat extensive coverage of Under-
    writers' Laboratories, but the extensive interaction of UL with
    other agencies seems to justify the attention I have paid to it
    here.)

## 12.11  Society of Automotive Engineers (SAE)

The Society of Automotive Engineers (SAE) was started in 1905 as the
Society of Automobile Engineers (see App. D for complete address).
Today it numbers among its members all kinds of engineering disciplines—mechanical, electrical, civil, chemical, aeronautical, physicists,
chemists, and scientists.   By its constitution, it is directed to "advance the Arts, Sciences, Standards, and Engineering practices,
principally concerned with self-propelled mechanisms . . .".   It is not
only an engineering society with more than 35,000 members, but it
develops and prints standards known as SAE standards.   These voluntary standards are listed in the SAE handbook.   Although they are
voluntary, many federal and state regulations reference these standards for minimum performance on safety items.

## 12.11.1  Standards Subscription Service

This SAE service provides engineers with up-to-date new and revised
standards and information reports, pertinent entries into the *Federal*

*Register*, and off-highway vehicle regulations as issued by organizations such as OSHA (Occupational Safety and Health Administration) and EPA (Environmental Protection Act).

### 12.11.2 Aerospace Standards

The *Aerospace Standards* (AS) are design or parts standards applicable to missiles, airline components, and accessory equipment. Also available are *Aerospace Recommended Practices* (ARP), and *Aerospace Information Reports* (AIR).

### 12.11.3 *Aerospace Material Specifications*

The *Aerospace Material Specifications* (AMS) cover materials, tolerances, quality control procedures, and processes. They also list chemical composition and detailed technical requirements, and provide for cross-reference of hundreds of similar material specifications on practically every material, including nonmetallic. More than 20 million individual specifications have been distributed.

The fact is, if it's mechanical and it moves, the SAE probably has a standard for it. For further information, write Society of Automotive Engineers, Inc. (See App. D for address.)

## 12.12 American Die Casting Institute

The American Die Casting Institute (ADCI) publishes a series of product standards for die castings (see App. D for complete address). They comprise four sections and are labeled accordingly: "E," Engineering; "M," Metallurgical; "C," Commercial; and "Q," Quality Standards. These standards have been prepared by ADCI as the representative of some 125 custom die-casting companies. Practically all casting drawings contain references to one or more of the ADCI standards. Write to American Die Casting Institute for information.

## 12.13 The Institute of Printed Circuits, Inc.

The Institute of Printed Circuits, Inc. (IPC) is a national trade association, founded in 1957 (see App. D for complete address). The institute developed from the need to exchange technical information on printed circuits, and for standards that would benefit all businesses associated with this field.

### 12.13.1 The Trade Association

This type of organization is typically American, and it is practically impossible to find a business field where there is not a trade association. Such an institution provides the involved businessman the opportunity to keep abreast of technological changes in the industry, and has provided documents by means of which the greatest efficiency and economy can be obtained. The activities of the trade association have contributed greatly to the American standard of living by insuring better quality products, generally at lower prices than would obtain if there were no focal point for the collection and dissemination of information on processes and practices.

## 12.14 The National Standards Association

At first glance, the National Standards Association (NSA) might be mistaken for the American National Standards Institute, but they are vastly different. The NSA was founded in 1946 (see App. D for complete address). It is principally a publishing and distributing organization. Its first published works were the National Aerospace Standards, but it soon expanded into many areas of both government and private standards.

### 12.14.1 Available Standards

Prominent among documents available from the NSA are National Aerospace standards (NAS), Air Force/Navy (AN), military standards (MS), Department of Defense (DOD), metric federal standards and specifications, military handbooks, military standards and specifications, and military and federal QPLs (qualified products lists), among others.

*12.14.1.1 Nonmilitary Standards.* Nonmilitary standards also available include those from ASTM, UL, NEMA, and the Radio Technical Commission for Aeronautics. These are available through what is called the DIMS (Direct Index Microfiche System).

(*Note*: More and more, documents are being provided on microfiche. Instead of printed pages, a subscription to a standards service brings flat film negatives of photographed documents. Receivers insert the film into a reader-printer and can make copies of whatever information they need. This system provides for a huge collection of documents in a very small area; one agency replaced 64 five-drawer steel filing cases with five files of microfiche, thus freeing valuable floor space and also making access to up-to-date documents much more effective. In my own standards office, we are currently beginning to replace many of our printed documents with film. Our principal concern is to have current documents for our engineers. The subscription services provide short interval updates, thus assuring current documents for engineering use and reference.)

(*Addendum*: This use of microfiche brings into focus the question of copyright infringement. Most of the standards organizations depend to a considerable extent on sale of their documents, and they cannot have them copied indiscriminately. Although there are laws protecting copyrights, they come into focus only after the act of infringement and during a legal proceeding. By and large, holders of copyright documents have to depend on inherent honesty among the users of their publications. There are different interpretations of what constitutes normal use and copying, and by no means do I want to pretend to be an authority on such usage. I believe for the purpose of this book it is sufficient to know that just because one has a microfiche does not mean he can indiscriminately make and sell prints.

The National Standards Association provides information on request.)

### 12.15 Department of Defense (DOD)

I had not intended to try to cover government standards sources, but the history of the Department of Defense is such that it has had a very great, and very important, impact on the standards used in America. There are about 40,000 documents. Following the Second World Ware there was a stark awareness of the liability of poor or inadequate standards, and many thousands of engineers, scientists, and technicians joined in an effort to provide the necessary documentation.

### 12.15.1 Application

These people became in the truest sense standards engineers. Their work provided the spacecraft that put Americans on the moon and supported the nuclear submarine development program. The body of standards and specifications became known and copied worldwide and was the basis for practically all industrial standards. In large measure, military standards and specifications still heavily influence industrial standards.

### 12.15.2 New Directions

Today governmental bureaus depend more and more on contracted standards, and even more on standards provided by the voluntary standards-writing associations. In 1976 DOD issued a document (Directive 4120.20) that mandated the use of nongovernment standards. This meant DOD people would actively participate in industrial standards-writing bodies and committees. The DOD further participated in writing what is known as OMB A-119 (Office of Management and Budget) circular.

The fact that government agencies will be more involved with and use standards developed by voluntary standards organizations also carries with it the obligation on the part of government, for documents involving national defense, to ensure the suitability of such documents.

Now I am going to say that in this book, and for the purpose of a general introduction to standards, I cannot give the DOD-ASTM interface/interaction/cooperation the space that would be required for a reasonable coverage of this very important situation. The reader is urged to contact ASTM directly for information and/or guidance concerning specific activities and subjects.

*12.15.2.1  Scope of OMBA-119.*  As this is written, OMB A-119 has not been formally accepted. This directive does much more than just say DOD people are to associate in voluntary standards. It provides that if these bodies want DOD participation, and if they want their standards to be used by the government, they have to meet government criteria, primarily associated with the philosophy of due process, whereby public citizen John Doe is protected from discrimination and injury as the result of processes that are not made public. In effect, the government will not do business with any standards organization that does not measure up to "standards" which it sees as protective of due process. For the main part, the various standards associations favor the OMB directive, especially when it is compared with efforts of the Federal Trade Commission to practically control the entire standards-making process in the public sector.

*12.15.2.2  Internationalization.*  There is still much that needs to be done to provide the nation with a strong standards establishment that can operate effectively both at home and abroad, but earnest efforts are being made to put a system together. If we don't, we will find America playing second fiddle on the international scene. (See paragraph 7.2.3.)

## 12.16  GATT Standards Code

The General Agreement on Tariffs and Trade (GATT) Standards Code, is about the most important agreement put together by a group of nations to ensure that the standards produced by one nation, or a bloc of nations, do not act adversely to restrict trade rights and opportunities of other nations. In effect, the GATT standards code applies the due process philosophy to the international scene. The agreement came into force on January 1, 1980. (See App. B.)

Article 10 concerns the provision of information about standards, technical regulations, and certification systems. It would require a separate document to fully present all aspects of the agreement, but one important part is that each signatory nation shall have what is designated as an "inquiry point" which is able to answer outside inquiries about national documentation which may be in process or in effect in the individual countries. The NBS serves this function for the United States. Similar inquiry points in other countries provide for an exchange of information about new or proposed changes to standards, technical regulations, and certification systems that is intended to provide any interested party an opportunity to comment on such.

This provision of an inquiry point practically forces a nation to operate its standards program on a national basis. This is not very difficult for most nations which are much smaller than the United States and have more authoritarian forms of government. In this country, however, you might very well wonder how any person, or group, could possibly know what is taking place in all the voluntary standards organizations in the country so as to be able to provide answers to other nations. Standards leaders in the United States are very well aware of the problems as well as the trade opportunities presented by GATT, and much work is going on to build an effective standards system. The years ahead are going to see much change and adaptation on the standards scene.

## 12.17 American Iron and Steel Institute

The American Iron and Steel Institute (AISI) has far more importance in materials than the brief coverage provided here (see App. D for complete address). Anyone involved with steel products should have a copy of the *Steel Products Manual* covering stainless and heat-resistant steels.

### 12.17.1 The Steel Products Manual

AISI is not a material specification-writing body, but some sections of the manual identify steels as "AISI standard." This designation refers to the chemical composition ranges and limits of such grades of steel. The manual is primarily concerned with manufacturing and testing of steel products, product properties and uses, dimensional tolerances, weight, inspection, sampling, and chemical analysis. Standards and specifications prepared by others are included. The manual includes equivalents in SI units of measurement (the modern metric system); since these are calculated equivalents to U.S. customery units, they may be approximate values. Contact the AISI for information.

## 12.18 National Aeronautics and Space Administration (NASA)

I had some doubts about including coverage here on the National Aeronautics and Space Administration (NASA) as it is not a standards body. However, the publications of this organization can provide current information for the engineer and it is well to know about its services (see App. D for complete address).

### 12.18.1 NASA Technology Utilization Program

The objective of this program, established in 1962, is to transfer to the public sector the benefits of aerospace research. Thousands of new ideas and applications have been made available to the public in potential products, shop and laboratory techniques, computer software, new concepts of the space age, and new sources of technical data.

*12.18.1.1 NASA Technical Briefs.* This is a quarterly publication, furnished free to engineers in U.S. industry. It presents a large number of brief descriptions about new developments in a wide variety of disciplines, and engineers, users, and manufacturers can find within its pages something new, informative, or challenging for further research and development. Many of the ideas are available for public patent. (See App. D for complete address.)

## 13. THE STANDARDS ENGINEERING SOCIETY (SES)

The Standards Engineering Society (SES) was organized in 1947 for the purposes listed below.

### 13.1 Provide an Association for Standards Engineers

The idea was that an association of standards engineers and other persons involved with standards would provide a means by which these people could, at meetings or through the publications of the society, discuss strategies of standardization considered to be of mutual benefit and interest.

### 13.2 Benefit the National Economy

The society was formed to further standardization as a means of advancing the national economy, and to promote a knowledge of the techniques and application of standards issued by regularly constituted standards bodies.

### 13.3 Production of Standards

The production of standards was definitely not to be a part of the activities of the Standards Engineering Society. The objective was not to create standards, but to bring together standards engineers where their total influence would work to the benefit of standardization and the best interests of everyone involved with or affected by standards.

### 13.4 Activities

Like any technical association, the SES desired to promote good fellowship among its members, based on their individual ideas and experiences. Regular meetings were established as a means of bringing the members together.

## 13.5 Publications

The primary publication of the SES is *Standards Engineering*, a magazine devoted to articles descriptive of the purposes of standardization. The pages of the publication are an open forum for the discussion of standards problems and solutions.

### 13.5.1 Reports

The magazine also carries reports of the activities of the society and of its members, and of other standards organizations, so that public awareness is enhanced.

### 13.5.2 Proceedings

The society publishes proceedings of the annual conference, where pertinent papers on standards are presented, and such other publications as may from time to time be considered of current value.

### 13.5.3 Public Relations

The society makes news of its activities available to trade journals and other media.

## 13.6 Membership

Membership is open to persons who are engaged in standardization activities, or who are interested in furthering the benefits of standardization, and who qualify under the laws of the society.

## 13.7 Sections

There are sections in various parts of the United States, and others are being formed as interest in standards becomes more evident and knowledgeable.

### 13.7.1 Activities of Sections

Section activities are keyed to national objectives, but additionally, section members strive to bring standards to the attention of the media and to educational institutions so the awareness of standards is increased. One objective of this is to urge more universities to offer studies leading to a standards engineer degree.

### 13.7.2 Accreditation

At this time the Standards Engineering Society is establishing an accreditation program whereby qualified people, by virtue of formal study or practical experience in standardization, can be certified by the society as certified standards engineers (not to be confused with accreditation). Such certification is based on a substantial level of professionalism. It is expected that this accent on professional status will encourage students to consider engaging in this profession.

(*Note:* This coverage has been provided on the SES because of its unique character of furthering and improving the public knowledge and public education, in regard to standardization. As mentioned in paragraph 13.3, the society does not write standards. For additional information about its activities, contact the SES. See App. A and D for complete address.

# 14. STANDARDS IN THE ENGLISH-SPEAKING COUNTRIES

It may seem natural that countries with a similar language should seek to coordinate their standards. This was the case in what was called the ABCA (American, British, Canadian, Australian) Unification of Engineering Standards (UES). While this organization was discontinued in 1980, the following information is included for historical interest and as an example of international standards motivation.

## 14.1 Beginning of Standards Coordination

The ABCA program began in the middle of World War II as the ABC program. The first British representative, in order to take part in a conference, flew from England to Africa, to South America, to the United States. The ABC was a loosely coordinated structure until the early 1960s, when it was finally realized that in order to be effective the structure had to be given more substance.

## 14.2 Development of ABCA

In 1972, a statement of principles was developed by the joint steering committee and Australia became a member, thus finally completing the roster of English-speaking nations.

## 14.3 Purpose

"To further the strength and effectiveness, both economic and military, of the participating countries (by implementation of Engineering standards)."

## 14.4 Objective

The objective of the ABCA was to ensure that selected engineering standards of one country were in common with and capable of being

clearly understood by the other three so that users of standards in the
four countries, when receiving engineering data, would be able to interpret and utilize the data with a minimum of inconvenience and delay.

## 14.5 ABCA Armies Standardization Program

When American arms were sent to England during the First World War,
it was found that such items as ammunition and fasteners would not
fit English products. From this very unpleasant experience, it was
realized that nations which were naturally allies should have a common
technical structure. Such a structure is only possible with commonly
understood standards.

### 14.5.1 Basic Language—Abbreviations

In 1968, representatives of the armies of the ABCA countries met in
London to reach an agreement on what standards would form the basis
for use, understanding, and defining abbreviations used on engineering drawings by the four countries. The concept was that such an
agreement would facilitate the interchangeability of products made in
accordance with each nation's drawings.

### 14.5.2 Other Areas of Standards Cooperation

Representatives from the ABCA countries subsequently met at a series
of meetings dealing with engineering drawing methods and graphic information. The objective was not only for internation coordination,
but for internation solidity on international committees of similar nature.

(*Note 1*: The mechanism for carrying forward the objectives of the
ABCA consisted of various committees and councils, including a Joint
ABCA Unification Steering Committee. Since the ABCA meeting in
Australia in 1980 was canceled, there will be no further activity for
this group. The reason, primarily, is that many of these people also
sit on international standards committees dealing with the same subject;
hence there would be a possible duplication of effort.)

(*Note 2*: As a personal experience, I can state that while all members
of the ABCA spoke "English," they applied this language lucidly from
various points of view, and with some rather large gaps in comprehend-
ing divergent viewpoints. However, the sincerity of intent overrode
the deficiencies, and if they could not agree, they at least reached a
point of common disagreement, in itself a very real value and an aid
to understanding. It can be expected than understanding between people speaking different languages and requiring interpreters is an even
more difficult process.)

14.5.3 Pacific Area Standards Congress (PASC)

For decades, standardization on the American continent faced East, across the Atlantic. The United States and Canada had natal ties with the European continent, and as the tide of progress flowed Westward, trade tended to flow to the East. Within the United States, also, as the country developed toward the Pacific, industrial ties and associations were focused on the industrial area of the Eastern states.

The inclusion of Australia in the ABCA (American, British, Canadian, Australian) conference, and its direct interest and interaction for mutually understandable engineering standards, brought a new element into the standardization picture.

In 1972, there was a meeting in Honolulu of representatives from the American National Standards Institute (ANSI), the Standards Association of Australia (SAA), the Canadian Standards Association (CSA) (see App. D), Underwriter's Laboratories (UL), and the Japanese Standards Association (JSA), to plan and develop a program for a voluntary independent organization of Pacific area national standards organizations. Japan had earlier desired to become a member of ABCA, but the direct intent of the ABCA was not ideally suited for inclusion of a non-English-speaking country. Additionally, the ABCA was never a strongly organized standards body.

The 1972 meeting was followed by another in 1973 and was called the Pacific Area Standards Congress (PASC). A list of 18 Pacific rim nations was drawn up, and the United States and Canada, both with Pacific coastlines, were included. It was intended that this new organization would not compete with ISO (the International Standards Organization), but would strengthen it by the addition of new memebers.

During the subsequent years, meetings were held at various locations, and the membership of nations changed as the result of wars that resulted in the dissolution of some governments and the formation of others, but the organization itself continued to function. It has passed numerous resolutions related to the activities of ISO and IEC (International Electrotechnical Commission), and was active in the formation of the Code of Principles on Certification to be embodied in the GATT Code of Conduct on Standards and Certification (App. B). With the influence of PASC, world standards have come much closer to being global in extent.

I would like to remind the student of the complexity of the many standards activities constantly occurring. Like a pond full of ducks, change is the constant, and there is not only a possibility but a probability that some of the information in this text is not the very latest. The student is urged to make personal contact with standards agencies identified in this text if explicit information is needed. A booklet entitled *Pacific Area Standards Congress* is available from ANSI.

## 15. CANADIAN STANDARDS

Trade and commerce have long existed between the United States and Canada, our nearest English-speaking neighbor, and an understanding of each other's standards has been essential to a successful interchange of goods and services. While it is not possible to go into depth on Canadian standards, it seems pertinent to provide some general information.

### 15.1 Beginning

The Canadian Standards Association had its beginning in 1919. It was originally formed to meet the needs for establishing engineering standards and to coordinate the efforts of producers and users for the improvement and standardization of materials.

### 15.2 Development

Up until the 1970s, standards work developed in an essentially unstructured environment. There were no national priorities and no national standards.

#### 15.2.1 Canadian Government Purchasing Standards Committee

In 1934, a significant advance had been made in the Canadian standards outlook when the Canadian Government Purchasing Standards Committee was formed under the National Research Council. Its function was to prepare federal government purchasing standards outside the engineering field, which was the province of the Canadian Engineering Standards Association.

#### 15.2.2 Standards Council of Canada

In 1970, the Standards Council of Canada Act created the Standards Council of Canada and assigned to it the broad objective of developing a national standards system for Canada.

The reason for this move was the recognition of the need for rationalizing standards-writing activities to avoid duplication of effort, and to permit effective utilization of limited technical resources available for standards work.

(*Note*: As this is written, a move is underway in the United States to develop a national standards policy for better control and utilization of the American standards effort. The fact that many ANSI-approved standards are called "American national" may mislead some people into thinking the standards are officially national. See paragraph 7.2.3.)

## 15.3 Canadian Standards Today

Canadian standards have not only caught up with the rest of the world, but in many instances standards "made in Canada" have established precedents for the standards of other countries.

Canada cooperates closely with the United States on standards (with many Canadian people on U.S. committees and vice versa), with the other ABCA countries, and with the Economic Commission of Europe (the "Common Market"). In brief, Canada partakes of world standards activities in much the same way as the United States, with memberships in the International Electrotechnical Commission (IEC).

## 16.  REVIEW

Up to this point, quite a large area of standards endeavor has been covered and the student should have a broad overview of what standards are and how they are created, and an awareness of some of the organizations involved.  The student should be able to do some personal research and follow-up with such organizations as seem most pertinent to his interests, or his field of work.  Apps. A and D should make contacts fairly easy.  The following paragraphs are a general description of some of the people who work in standards, aside from committee members whose function has already been described.

### 16.1  The Standards Technician

The nature of this type of work depends to a large extent on what kind of standards endeavor the person is involved in.

### 16.1.1  Research

In some areas of standards, the technician may be involved with documentary research.  This means having the responsibility of securing facts and figures from existing documents in many different disciplines, generally as an assistant to the standards engineer.  The objective of such searches is to find out what has been written before on a subject on which a new standard is desired, or for revising an existing standard.  Without this sort of research, many similar standards would tend to duplicate or contradict each other, and in so doing would pretty well nullify the effectiveness of the whole.

(*Note*:  This type of research is becoming less tedious due to the fact that worldwide information networks, based on the computer, are coming into increasing use.  For a fee, depending on the subject, computers can provide information which either would not have been possible to secure, or which would have previously taken an intolerable length of time to collect.)

*16.1.1.1  Requirements for the Research Technician.*  The research technician must have an active and inquiring mind and the capacity for remembering subjects in different types of documents. An easy and familiar acquaintance with words and word usage is essential.

### 16.1.2 Experimental Technician

This form of standards work requires the capability of using laboratory equipment and of utilizing the scientific method to ascertain results of experiments. It is not necessary to be an expert in language, although there must be an ability to write clear reports.

## 16.2 The Standards Writer

People who undertake to write standards must have a comprehensive understanding of language and composition. It is necessary that they be able to visualize format and organize material (such as the notes or reports provided by research technicians), and express such material in a clearly understandable manner. This is not an easy thing to do as a standard is used by a great many people who may come from a variety of disciplines. It is not necessary for the writer, generally, to be an engineer—in fact, there may be some advantage in not having a specialized field. However, it is good if the writer has "knocked around" a bit and gained some familiarity with many different engineering and scientific fields.

## 16.3 The Standards Engineer

It may seem strange to readers of this text that I do not have a stylized description of a standards engineer. At one time I was told that two universities in America have a degree program for this subject, but I was not able to identify these schools. From personal experience, I find that most standards engineers are specialists in a given engineering field and are involved in their work with standards pertaining to that subject. Others are widely experienced people who have "grown up" in a daily atmosphere of standards work and developed expertise as they progressed. Direct requests for more information may be made to the Standards Engineering Society (paragraph 13).

## 17. NATIONAL CONFERENCE OF STANDARDS LABORATORIES (NCSL)

It has not been the intent of this text to cover the various types of testing laboratories, except for the National Bureau of Standards and the Underwriters' Laboratories, Inc. However, the National Conference of Standards Laboratories (NCSL) deserves some mention (see App. D for complete address).

### 17.1 Organization

The NCSL is an organization of professional standards people who operate standards laboratories for the purpose of testing, measurement, and calibration. For the most part, members are representatives of industrial laboratories. There are about 300 members.

### 17.2 Objectives

The goals of the conference are to provide an exchange of information between standards laboratories that will enable members to solve problems through an interchange of experiences related to calibration, particularly in the field of product design and specification.

### 17.3 Sponsor

While the membership of NCSL consists primarily of independent laboratories, its formation was sponsored by the National Bureau of Standards with the objective of maintaining a high degree of technical excellence in measurement and calibration throughout the United States. The main interactive body was/is the Institute of Basic Standards (IBS).

### 17.4 Benefits

As with any type of standard, the common application of calibration procedures, and the creation of specifications controlling these applications, are a valuable addition to the overall resources of the nation.

17.4.1 Developing Countries

Beyond the obvious fact of benefits to the nation, the common approach and application of measurement standards in this country lends to the capacity of assisting undeveloped countries in establishing individual or national standards laboratories. It is a good thing for everyone when confusion of measurement, whether between people or between nations, is minimized.

## 18. FACTORY MUTUAL SYSTEM (FM)

In our discussion of Underwriters' Laboratories, Inc. (paragraph 12.10), we may have given the reader the impression that UL was a "one and only" product-safety testing organization. Actually, there are many other organizations, including Factory Mutual, which is a very large organization (see App. D for complete address).

### 18.1 History  ·

More than a century ago (actually in 1635), a group of textile manufacturers in New England organized an insurance system to protect their resources. Primarily, this protection was against fire losses, but heavy emphasis was placed on fire protection, and the scientific study of causes. The system rapidly expanded into engineering and research.

### 18.2 Today

Today, FM provides many kinds of research, testing, and listing of products. It also provides consultative service, training, and, of course, fire insurance. It is one of the world's foremost insurance organizations.

### 18.2.1 Scope

The FM system will accept equipment, materials, and services for approval testing for safety and reliability in much the same way as UL. It is officially recognized by OSHA (Occupational Safety and Health Administration) as a fully accredited testing organization.

### 18.3 Approved Products

Factory Mutual issues an FM approval guide that lists over a thousand companies.
For more information, write to Factory Mutual System.

## 19. A TIME TO STOP

In the pages of this text, I have but brushed lightly over the subject of standards. If what I have said here stimulates the student to learn more, then the objective has been reached.

## APPENDIX A: INFORMATION SOURCES

ANSI American National Standards Institute, 1430 Broadway, New York, NY 10018

ANSI Reporter. Biweekly; articles on standards activities

Standards Action. Biweekly; information on new standards and revisions to standards; invites public comment

ASTM American Society for Testing and Materials, 1916 Race Street, Philadelphia, PA 19103

Standardization News. Articles and information on new standards; ballot forms for membership approval

CSA Canadian Standards Association, 178 Rexdale Boulevard, Rexdale, Ontario, M9W IR3 Canada

Standards Canada. Quarterly; standards articles

EIA Electronic Industries Association, 2001 Eye Street NW, Washington, DC 20006

Index of EIA and JEDEC (Joint Electron Devices Engineering Council) standards and publications

IEC International Electrotechnical Commission, 1, rue de Varembé, 1211 Geneva 20, Switzerland

Bulletin. Contains articles, new publications

IEEE Institute of Electrical and Electronic Engineering, Service Center, Publications Sales Department, 445 Hoes Lane, Piscataway, NJ 08854

Publications bulletin

ISA Instrument Society of America, 67 Alexander Drive, P.O. Box 12277, Research Triangle Park, NC 27709

Instrument Technology. Official publication of the Society; covers all aspects of measurement and control systems in industry, science, and research, including computer-based systems. Also publishes English translations of Russian journals

ISO International Standards Organization, 1, rue de Varembé, 1211 Geneva 20, Switzerland

ISO Bulletin. Information on international standards

NBS National Bureau of Standards, *NBS Publications Newsletter*, Technical Information and Publications Division, Washington, DC 20234
  *NBS Publications Newsletter*. Contains information on new publications
NEMA National Electrical Manufacturers Association, 2101 L Street NW, Washington, DC 20037
  NEMA standards publications
SAA Standards Association of Australia, Standards House, 8086 Arthur Street, North Sydney, Australia 2060
  *Australian Standards*. Monthly journal of the Standards Association of Australia; lists new standards and some articles
SES Standards Engineering Society, 6700 Penn Avenue South, Minneapolis, MN 55423
  *Standards Engineering*. Articles on standards
UL Underwriters' Laboratories, UL Publications Stock, 333 Pfingston Road, Northbrook, IL 60062
  *Lab Data*. Quarterly technical and information publication
  Catalog

The organizations on the foregoing list are actively engaged with the development and use of standards. Now I would like to also include the name of an organization which has the primary objective of providing current news about standards and activities concerning standards. This is the Marley Organization, Inc., Resources and Information Services, 11 Todds Road, Ridgefield, CT 06877.

This company publishes a twice-monthly report concerning developments and affairs in national and international standards, certification, and accreditation.

While the report is a subjective analysis of events, and is written in a rather staccato style, it provides a wealth of information to subscribers based on many years of experience by the author in standardization and related activities.

The Marley Organization can also provide lectures, seminars, and so on, on standards.

## APPENDIX B: THE GENERAL AGREEMENT ON TARIFFS AND TRADE

In regard to international standards, in April 1979, negotiations between nations that had been underway for many years reached an agreement that has come to be called the "Standards Code" and is a part of the General Agreement on Tariffs and Trade (GATT) code. The agreement was signed after Congressional hearings in July 1979, and is known as the Trade Agreements Act of 1979.

What led to the creation of GATT was the realization among nations that it was possible for any nation, or bloc of nations, to unfairly restrict trade (in their own benefit) by means of standards. It was further realized that in world trade it was no longer tenable that products of one nation should not harmonize (receive equal treatment) with those of other nations.

The Standards Code requires signatory nations to abide by certain practices, which may be identified as follows: (1) national standards will be based on international standards; (2) international standards will be used if such exist and are applicable; (3) when nations develop their own standards they will do so openly and will publicize the fact for other nations' awareness; (4) there will be mutual acceptance of other nations' test methods and certification. Additionally, performance criteria (how a product will act, and what it will do) instead of design criteria (how it is to be made) will be used to the fullest extent possible. Each nation will also encourage participation in international standards-making bodies and will provide access to its national and regional certification systems.

The student will again see the need for an American national standards policy. Standards between nations are no longer for amateurs; in fact, we may yet see a federal cabinet position similar to the Soviet Union's Minister of Standards.

This brief review of the GATT code can only touch the key points of an agreement that is actually a very complex document. I suggest contacting the Special Trade Representative, the White House, Washington, DC, if further information is desired.

# APPENDIX C: INSTRUMENT SOCIETY OF AMERICA
## SAMPLE BALLOT

The following information is included in this book by courtesy of the Instrument Society of America (see App. D for complete address). It is an excellent example of an effort to secure a consensus approval for a document.

### INSTRUMENT SOCIETY OF AMERICA

To: Review Board for ISA Draft Standard dS12.12.

The attached draft dated August 1981 was prepared by ISA Committee SP12.12 as part of ISA's continuing effort to provide useful voluntary standards in the instrumentation field for science and industry. If comments were received to earlier drafts, such comments were carefully evaluated by the committee and suitable revisions made as reflected in the attached document.

Acceptance of this standard is being sought in order to develop evidence of consensus on which approval as an American National Standard will be based.

We respectfully request that, after familiarizing yourself with the scope of this document, you review all portions of this draft, denote your judgment on the mail ballot, and return the ballot to us. Ballot due date is _____.

The ballot is accompanied by instructions so the responder can feel at ease and knowledgeable about what his "vote" means. These instructions are reproduced on the following pages.

## MAIL BALLOT

---

ISA standard S12.12          Draft no. 10                    Dated 1981

("A") Approved

("D") Approved, but would like my comments (on attached sheet)
    considered.

("M") Approved, provided that my comments (on attached sheet)
    are followed.

("X") Abstain

("N") Disapproved for the following reason(s) _____

_____

_____

_____

Name _____ Position _____

Business address _____

               _____

               _____

Industry _____ Years experience in this field _____

Classification: Producer interests                    (P)
              Distribution or retailer interests    (D)
              Consumer or user interests            (U)
              General interests                     (G)

---

**Figure 1** Instrument Society of America sample mail ballot.

## INSTRUCTIONS TO REVIEWERS OF dS12.12

You have offered to serve on a board of review of the attached draft standard. A ballot and a form for making comments on this standard are attached. The ballot must be used to record your vote on this draft standard and the comments form is provided to give the committee an accurate record of all comments, but any comments will be considered by the committee even if the form is not used. In particular typographical corrections or errors in diagrams are frequently best recorded by marking up the draft standard. Please make as many copies of the review form as you need. It is preferable to use a separate sheet for each major comment, especially substantive comments that are the basis of a ballot for disapproval.

As a reviewer you should be aware of the significance of your ballot. The standards committee and the Standards and Practices Board (S & P) of the ISA will interpret your vote as follows:

Abstain (X). You decline to give an opinion on the standard usually because you are not knowledgeable in this area. The committee will note the number who abstain but these votes have no impact on the standard review process.

Approve (A). You approve the standard as written without comment. The committee will assume that this approval will continue even if minor changes are made in the standard.

Approve with nonmandatory comments (D). You approve the standard but have some comments you wish the committee to consider. If the comments are not included in the standard you still approve the standard. The committee will reply to your comments and if minor changes are made in the standard will assume your approval continues.

Approve with mandatory comments (M). You approve the standard provided your comments are included in the standard. If the comments are not included you disapprove the standard. The committee will reply to your comments and ask you to change your vote to "A" (approve) as a result of changes to the standard or provide substantive reasons why the committee does not agree with your comments. Failure to reply to the committee's comments in a reasonable time will be considered to indicate approval.

Disapprove (N). You disapprove the standard. A disapproval without reason is of no assistance to the committee or the S&P Board. The committee will reply to your comments and ask you to change your vote to "A" (approve) as a result of changes to the standard or provide substantive reasons why the committee does not agree with your comments. Failure to reply to the committee's comments in a reasonable time will be considered to indicate approval.

In making comments, constructive suggestions are encouraged and proposed additions or changes are welcomed but you should include the reason for the change. Purely negative comments such as "you should not be doing this work" or "this is not the way I like to do it" are not helpful and are unlikely to receive much attention by the committee. The recommended changes should be specific and, if possible, written so they can be easily incorporated into the standard.

The ISA Standards and Practices Board requires that all standards are reviewed by a board representative of all groups interested in the scope of the standard. To ensure that all parties are adequately represented in the Board of Review you are asked to indicate your classification. This classification refers to you and your present occupation or expertise used in evaluating the standard. It does not necessarily refer to the classification of your company. The classifications are defined as follows:

1. Producer interests (P). Those directly concerned with the product itself. This classification includes personnel who are involved in the design, engineering support, manufacturing, testing, and marketing of a product and who are employed by the manufacturer (producer) of the product. When a producer's employees are involved with a standard covering a product purchased by their employers, they should classify themselves as users.
2. Distribution and retailer interests (D). Those independently concerned (not associated with a producer) with the marketing of the product between producer and consumer.
3. Consumer or user interests (U). Those who use the product involved but are not involved with its production or distribution. In addition to representatives of the end user or consumer, this classification also includes representatives of engineer-constructor and architect-engineer organizations who are involved with the application, installation, and use of the product.
4. General interest (G). Those who have interests other than those described in (1) through (3) (e.g., government agencies, educators, consultants, etc.).

## APPENDIX D: ABBREVIATIONS AND ACRONYMS

The abbreviations and acronyms in the following list were gathered from many public sources, including the *Federal Register* and the National Bureau of Standards. A study of the list can be rewarding, not only for the information it contains, but as an aid to further understanding of the vast field encompassed by standards.

Not all abbreviations identify organizations. For those that do, I have either provided addresses, or made reference to other organizations where information can be obtained. I recommend the reader consult the following: Public Library; *Encyclopedia of Associations*, Gale Research Company, Book Tower, Detroit, Michigan 48226; *National Trade and Professional Associations* (NTPA), Columbia Books, Inc., Publishers, 777 14th Street, NW, Washington, DC 20005; the American National Standards Institute (ANSI, address given below); U.S. Government Printing Office, Washington, DC 20402.

| | |
|---|---|
| AA | The Aluminum Association, 818 Connecticut Avenue NW, Washington, DC 20006 |
| AAAS | American Association for the Advancement of Science, 1515 Massachusetts Avenue NW, Washington, DC 20005 |
| AALA | American Association for Laboratory Accreditation (See UL) |
| AAMI | Association for the Advancement of Medical Instrumentation, 1901 North Fort Myer Drive, Suite 602, Arlington, VA 22209 |
| AATT | American Association for Textile Technology, Inc., 1040 Avenue of the Americas, New York, NY 10018 |
| ACI | American Concrete Institute, P.O. Box 19150, Redford Station, Detroit, MI 48219 |
| ACIL | American Council of Independent Laboratories, 1725 K Street NW, Washington, DC 20006 |
| ADA | American Dental Association, 211 East Chicago Avenue, Chicago, IL 60611 |

| ADCI | American Die Casting Institute, 366 Madison Avenue, New York, NY 10012 |
| AES | Audio Engineering Society, Inc., 60 East 42nd Street, New York, NY 10017 |
| AFBMA | Anti-Friction Bearing Manufacturers Association, 2341 Jefferson Davis Highway, Arlington, VA 22202 |
| AFIPS | American Federation of Information Processing Societies, 1815 North Lynn Street, Arlington, VA 22209 |
| AGA | American Gas Association, 1515 Wilson Boulevard, Arlington, VA 22209 |
| AGMA | American Gear Manufacturers Association, Suite 1000, 1901 North Fort Myer Drive, Arlington, VA 22209 |
| AIAA | American Institute of Aeronautics and Astronautics, 1290 Avenue of the Americas, New York, NY 10019 |
| AHAM | Association of Home Appliance Manufacturers, 20 North Wacker Drive, Chicago, IL 60606 |
| AIA | Aerospace Industries Association, 1725 De Sales Street NW, Washington, DC 20036 |
| AIChE | American Institute of Chemical Engineers, 345 East 47th Street, New York, NY 10017 |
| AIMME | American Institute of Mining, Metallurgical, and Petroleum Engineers, 345 East 47th Street, New York, NY 10017 |
| AISC | American Institute of Steel Construction, 400 North Michigan Avenue, 8th Floor, Chicago, IL 60611 |
| AISE | Association of Iron and Steel Engineers, 3 Gateway Center, Suite 2350, Pittsburgh, PA 15222 |
| AISI | American Iron and Steel Institute, 1000 16th Street NW, Washington, DC 20036 |
| ALA | American Library Association, 50 SE Huron Street, Chicago, IL 60601 |
| AMA | American Medical Association, 535 South Dearborn Street, Chicago, IL 60601 |
| AMS | Aerospace Material Specifications (See NSA) |
| AMSA | American Metal Stamping Association, 27027 Charlton Road, Richmond Heights, Ohio 44143 |
| ANMC | American National Metric Council, 1625 Massachusetts Avenue NW, Washington, DC 20036 |
| ANS | American Nuclear Society, 555 North Kensington Avenue, LaGrange Park, IL 60525 |
| ANSI | American National Standards Institute, 1430 Broadway, New York, NY 10018 |
| AN-USM | Army Navy/U.S. Military (See NSA) |
| API | American Petroleum Institute, 2101 L Street NW, Washington, DC 20037 |

| | |
|---|---|
| APICS | American Production and Inventory Control Society, Suite 504, 2600 Virginia Avenue NW, Washington, DC 20037 |
| ARI | Air Conditioning and Refrigeration Institute, 1815 North Fort Myer Drive, Arlington, VA 22209 |
| ASCE | American Society of Civil Engineers, 345 East 47th Street, New York, NY 10017 |
| ASCII | American National Standard Code for Information Exchange (See NSA) |
| ASEE | American Society for Engineering Education, One Dupont Circle NW, Suite 400, Washington, DC 20036 |
| ASHRAE | American Society for Heating, Refrigeration, and Air Conditioning Engineers, 345 East 47th Street, New York, NY 10017 |
| ASM | American Society for Metals, Metals Park, OH 44073 |
| ASME | American Society of Mechanical Engineers, 345 East 47th Street, New York, NY 10017 |
| ASNT | American Society for Nondestructive Testing, Inc., 3200 Riverside Drive, Box 5642, Columbus, OH 43221 |
| ASQC | American Society of Quality Control, Inc., 161 West Wisconsin Avenue, Milwaukee, WI 53203 |
| ASTM | American Society for Testing and Materials, 1916 Race Street, Philadelphia, PA 19103 |
| AWS | American Welding Society, 2501 NW 7th Street, Miami, FL 33125 |
| CBEMA | Computer and Business Equipment Manufacturers Association, 1828 L Street NW, Washington, DC 20036 |
| BIPM | International Bureau of Weights and Measures, Pavilon de Breteuil, F-92310, Sevres, France |
| BSI | British Standards Institute, 2 Park Street, London, W1 A2BS England |
| CEE | International Commission on Rules for the Approval of Electronic Equipment, Utrechseweg 310, Arnhem, Netherlands |
| CEMA | Canadian Electrical Manufacturers Association (See EEMAC) |
| CEN | European Committee for Standardization (See ISO) |
| CENELEC | European Committee for Electrotechnical Standardization (See ISO) |
| CERTICO | Committee on Certification (ISO) |
| CGPM | General Conference on Weights and Measures (See ISO) |
| CIE | International Commission on Illumination (See ISO) |
| CIPM | International Committee for Weights and Measures (See ISO) |
| CIRP | International Institution for Production Engineering Research (See ISO) |

| | |
|---|---|
| COPANT | Pan American Standards Commission (See ANSI) |
| CSA | Canadian Standards Association, 178 Rexdale Boulevard, Ontario, M9W 1R3, Canada |
| CSC | Canada Safety Council (See CSA) |
| DIN | Deutsches Institut for Normung (German Standards Institute), Beuth Verlag GmBH, Berlin, Germany KoLN |
| DOD | Department of Defense, Naval Publications and Forms Center, 5801 Tabor Avenue, Philadelphia, PA 19120 |
| DODISS | Department of Defense Index of Specifications and Standards (See DOD) |
| ECMA | European Computer Manufacturers Association, Rue du Rhone 114, 1204 Geneva, Switzerland |
| EEC | European Economic Community, 2100 M Street, NW, Suite 707, Washington, DC 20037 |
| EEMAC | Electrical and Electronic Manufacturers Association of Canada, Yonge Street, Suite 1608, Toronto, Ontario, M5E 1R1 Canada |
| EPTA | European Free Trade Association (See ISO) |
| EIA | Electronic Industries Association, 2001 Eye Street, NW, Washington, DC 20006 |
| EPA | Environmental Protection Agency (See NBS) |
| FASST | Forum for the Advancement of Students in Science and Technology, Inc., 1785 Massachusetts Avenue NW, Washington, DC 20036 |
| FCC | Federal Communications Commission (See NBS) |
| FDA | Food and Drug Administration (See NBS) |
| FIPS | Federal Information Processing Standards (See NBS) |
| FM | Factory Mutual Engineering Corporation, Factory Mutual System, 1151 Boston-Providence Turnpike, Norwood, MA 02062 |
| FTC | Federal Trade Commission (See NBS) |
| GIDEP | Government/Industry Data Exchange Program (See NSA) |
| IAEA | International Atomic Energy Agency, Vienna International Center, P.O. Box 100, Wagramerstrasse 5 A-1400, Vienna, Austria |
| IAESTE | International Association for the Exchange of Students for Technical Experience, American City Building, Suite 217, Columbia, MD 21044 |
| ICRM | International Commission on Radiation Units and Measurements, 7910 Woodmont Avenue, Suite 1016, Washington, DC 20014 |
| ICRP | International Commission on Radiological Protection (See ISO) |
| IEC | International Electrotechnical Commission, 1, rue de Varembe, 1211 Geneva 20, Switzerland |

| | |
|---|---|
| IEEE | Institute of Electrical and Electronic Engineers, 345 East 47th Street, New York, NY 10017 |
| IFAN | International Federation for the Application of Standards (See ISO) |
| ILAC | International Laboratory Accreditation Conference (See ANSI) |
| IPC | Institute for Interconnecting and Packaging Electronic Circuits, 3451 Church Street, Evanston, IL 60203 |
| ISA | Instrument Society of America, 67 Alexander Drive, P.O. Box 12777, Research Triangle Park, NC 27709 |
| ISO | International Organization for Standardization, 1, rue de Varembe, 1211 Geneva 20, Switzerland |
| ISONET | ISO Information Network (See ISO) |
| ITI | International Technical Information Institute, Toranomon-Tachikawa Building, 165 Nish-Gijutsu, Minato Ku, Tokyo 105, Japan |
| ITU | International Telecommunications Union, Place des Nations, 1211 Geneva 20, Switzerland |
| JEDEC | Joint Electron Device Engineering Council, 2001 Eye Street, Washington, DC 20006 |
| JIMS | Joint Industry-Military Standard (See NSA) |
| NAF | Netherlands-American Foundation, c/o Netherlands Consulate, One Rockefeller Plaza, New York, NY 10020 |
| NASA | National Aeronautics and Space Administration, Technology Transfer Division, P.O. Box 8757, Baltimore/Washington International Airport, Maryland 21240 |
| NBS | National Bureau of Standards, U.S. Department of Commerce, NBS, Washington, DC 20234 |
| NCSL | National Conference of Standards Laboratories, National Bureau of Standards, Room 4001 Radio Building, Boulder, CO 80303 |
| NCWM | National Conference on Weights and Measures, National Bureau of Standards, Washington, DC 20234 |
| NEMA | National Electrical Manufacturers Association, 2101 L Street NW, Washington, DC 20037 |
| NFPA | National Fire Protection Association, 470 Atlantic Avenue, Boston, MA 02210 |
| NFP(A) | National Fluid Power Association, 3333 North Mayfair Road, Suite 311, Milwaukee, WI 53222 |
| NSA | National Standards Association, 4827 Rugby Avenue, Washington, DC 20014 |
| NSC | National Safety Council, 444 North Michigan Avenue, Chicago, IL 60601 |
| NSPAC | National Standards Policy Advisory Committee, Suite 935, Washington Building, Washington, DC 20005 |

NTIS            National Technical Information Service, United States
                Department of Commerce, Springfield, VA 22161
NVLAP           National Voluntary Laboratory Accreditation Program
                (See NBS)
OIML            International Organization for Legal Metrology (See
                ISO/ANSI)
OMB             Office of Management and Budget, Washington,
                DC 20503
OSHA            Occupational Safety and Health Administration, United
                States Department of Labor, Washington, DC 20210
PASC            Pacific Area Standards Congress (See ASNI)
SAA             Standards Association of Australia, Standards House,
                80-86 Arthur Street, North Sydney, New South Wales,
                Australia
SAE             Society of Automotive Engineers, 400 Commonwealth
                Drive, Warrendale, PA 15096
SAMA            Scientific Apparatus Makers Association, Suite 300,
                1101 16th Street NW, Washington, DC 20036
SCC             Standards Council of Canada, 350 Sparks Street,
                Ottawa, Ontario K1R 758 Canada
SES             Standards Engineering Society, 6700 Penn Avenue,
                South, Minneapolis, MN 55423
SIS             Scientific Information Service (See ANSI)
STC             Society for Technical Communication, Suite 506, 815
                15th Street SW, Washington, DC 20005
UL              Underwriters Laboratories, Inc., Public Information
                Office, 333 Pfingsten Road, Northbrook, IL 60062
USMA            United States Metric Association, Sugarloaf Star Route,
                Boulder, CO 80302

(*Note*: As previously stated, the objective in providing this list is to
give a sampling of organizations dealing with standards. There are
thousands of such, and the references at the beginning of this index
will provide further identities.)

# INDEX

*Numbers refer to paragraphs.*